VI Workshop in Electron Volt Neutron Spectroscopy: Frontiers and Horizons 2014

Journal of Physics: Conference Series Volume 571

Abingdon, United Kingdom
20 – 21 January 2014

ISBN: 978-1-63439-756-8
ISSN: 1742-6588

VI Workshop in Electron Volt Neutron Spectroscopy: Frontiers and Horizons 2014

Journal of Physics: Conference Series Volume 571

Abingdon, United Kingdom
20 – 21 January 2014

ISBN: 978-1-63439-756-8
ISSN: 1742-6588

IOP Institute of Physics

Printed from e-media with permission by:

Curran Associates, Inc.
57 Morehouse Lane
Red Hook, NY 12571

Printed by Curran Associates, Inc. (2015)

For permission requests, please contact the Institute of Physics
at the address below.

Institute of Physics
Dirac House, Temple Back
Bristol BS1 6BE UK

Phone: 44 1 17 929 7481
Fax: 44 1 17 920 0979

techtracking@iop.org

Additional copies of this publication are available from:

Curran Associates, Inc.
57 Morehouse Lane
Red Hook, NY 12571 USA
Phone: 845-758-0400
Fax: 845-758-2634
Email: curran@proceedings.com
Web: www.proceedings.com

TABLE OF CONTENTS

Journal of Physics: Conference Series **571** (2014) 011001 doi:10.1088/1742-6596/571/1/011001

The VI Workshop in Electron Volt Neutron Spectroscopy: Frontiers and Horizons

A.G. Seel,[1] R. Senesi,[2,3] and F. Fernandez-Alonso[1,4]

[1] ISIS Facility, Rutherford Appleton Laboratory, Chilton, Didcot, Oxfordshire OX11 0QX, United Kingdom
[2] Università degli Studi di Roma Tor Vegata, Dipartimento di Fisica and Centro NAST, via della Ricerca Scientifica 1, Rome 00133, Italy
[3] CNR-IPCF Sezione di Messina, Messina, Italy
[4] Department of Physics and Astronomy, University College London, Gower Street, London WC1E 6BT, United Kingdom

January 2014 saw the congregation in Abingdon (United Kingdom) of scientists from across the world, to discuss the current state and future of spectroscopy using epithermal neutrons. This meeting was the sixth in a series of workshops held in collaboration between the Science and Technology Facilities Council (United Kingdom) and the Consiglio Nazionale delle Ricerche (Italy), aimed at bringing together researchers with an interest in the use of electron-volt neutrons in spectroscopic studies [1]. This technique is termed Deep Inelastic Neutron Scattering (DINS), and also Neutron Compton Scattering (NCS) in reference to the analogy with Compton scattering of X-rays from electrons. In particular, this meeting centred jointly around experimentalists and theoreticians, formulating animated discussions as to the current overview of DINS and future horizons facing the field.

The use of electron Volt neutrons for spectroscopic measurements dates back to the advent of proton-driven spallation neutron sources in the 1970s and 1980s. Following an initial scientific meeting in Los Alamos (USA) [2], the first two meetings in this series were held in Abingdon (United Kingdom) in 1995 and 1998, with subsequent meetings held in Santa Fe (USA) in 2005 [3], Oak Ridge National Laboratory (USA) in 2006 [4], and Rome (Italy) in 2010 [5]. The flagship instrument serving a continual user-programme for DINS measurements, and the main focus of this meeting, has been the VESUVIO spectrometer at ISIS [6, 7]. Subsequent instruments like SEQUOIA in the USA [8, 9] and a newer spectrometer in the Bariloche LINAC in Argentina [10, 11] have also been commissioned and began serving a growing user community.

The availability of DINS measurements has extended the range of possible spectroscopic techniques that utilise neutrons into the kinematic region of high energy and momentum transfers, shown schematically in Figure 1. Spectroscopic instrument suites such as that of ISIS are thus able to probe processes on timescales across nine orders of magnitude, from quantum tunnelling and molecular diffusion up to nuclear recoil under the impulse approximation.

DINS is one of the most direct techniques for probing nuclear quantum dynamics in condensed matter, with the ability to measure nuclear kinetic energies and momentum distribu-

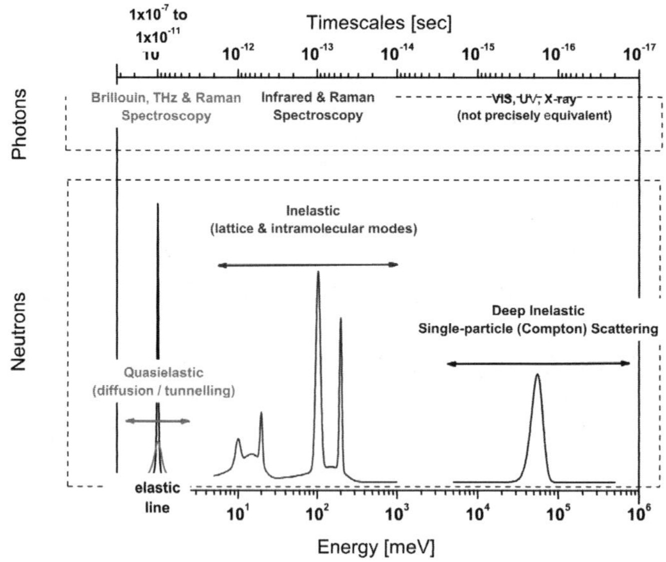

Figure 1. Energies and associated timescales accessible with neutron spectroscopy. The DINS regime remains the exclusive realm of accelerator-driven pulsed neutron sources.

tions of light nuclei. A major theme for current DINS research relates to the study of nuclear quantum effects, particularly regarding the non-classical behaviour of hydrogen or deuterium. This emphasis was reflected by the meeting's opening session dealing with the measurement of momentum distributions in hydrogen-bonded and aqueous systems (see Table 1). Whilst concentrating on experimental studies, data were presented from the outset alongside state-of-the-art path-integral molecular dynamics (PIMD) calculations, affording an animated discussion as to the interpretation of measured neutron Compton profiles in spatially disordered systems.

The extension of DINS to heavier nuclei was visited in the second session of the meeting, under the over-arching theme of MAss-selective Neutron SpEctroscopy (MANSE). The ability of DINS to examine atomically distinct species in condensed matter has garnered interest from the chemistry and materials communities, with both binary systems and more complex materials being presented in this session. Nuclei of interest in applied materials research were highlighted, including oxygen, lithium and fluorine, alongside an assessment of current capabilities and future developments in their examination by DINS.

The concluding session of the first day of the meeting was concerned with instrumentation. The detection of epithermal neutrons requires technologies unique relative to their lower-energy counterparts, and the first talk of this session outlined how detector capabilities on existing instruments has been revolutionised in recent years. Resonance detectors and techniques involving analyser-foil cycling were detailed, explaining how they have improved not only the count-rate but also the spectral resolution of instruments such as VESUVIO. Beyond NCS, the detection of nuclear resonances were also considered. An overview of VESUVIO (a time-of-flight instrument operating in so-called indirect geometry) was then given, alongside ways it could be improved upon in light of growing demands from the user community. This discussion was complemented

Figure 2. Concluding photograph for the *VI Workshop in Electron Volt Neutron Spectroscopy: Frontiers and Horizons*, January 21st, 2014.

by a presentation of the use of direct-geometry instruments such as SEQUOIA for DINS measurements, highlighting the complementarity with parallel studies on VESUVIO.

The second day of the meeting was given over to the theoretical aspects of nuclear momentum distributions. Delegates were presented with the manner in which we can computationally incorporate nuclear quantum effects, along with their output in terms of 'measureables,' relating to a varied range of systems. Results were presented from hydrogen-bonded liquids, ferroelectric materials, biological systems, and mixed bosonic-fermionic calculations. The ability to now approach experimental measurement with theoretical insight, and vice versa, is a sea-change in research into nuclear quantum effects. This field is a high-profile area in chemical physics and materials research, and a concerted approach by theory and experimentation allows us to explore and explain the dynamical state of matter at the atomic level in a manner previously unobtainable.

Concluding each session throughout the meeting were periods of open discussion, wherein all present could comment or expand upon themes arising from the preceding talks. These discussions were then further distilled and clarified during the final section of the meeting, the 'Frontiers and Horizons' discussion, open to all present to delineate the next steps in DINS as a whole. Experimentalists and theoreticians were able to present their views on how each community could move forward to complement each other, not only separately but with specific projects and milestones in mind. These discussions have been included within these proceedings to highlight the future outlook of DINS.

It is the hope of these editors that this meeting will allow spectroscopy with epithermal neutrons to further develop, explaining the very quantum nature of dynamics in condensed matter. Bringing together, as it did, international researchers in fields at the frontiers of their respective experimental and theoretical communities, we hope that the next meeting in this series will see the fruition of concepts and projects instigated by this one.

Journal of Physics: Conference Series **571** (2014) 011001 doi:10.1088/1742-6596/571/1/011001

List of Registered Attendees

Prof. Carla Andreani, Università di Roma - Tor Vergata, Italy
Prof. Francisco Javier Bermejo, Consejo Superior de Investigaciones Científicas, Spain
Prof. Fernando Bresme, Imperial College London, United Kingdom
Mr. Peter Byrne, Durham University, United Kingdom
Dr. Carlos Cabrillo, Consejo Superior de Investigaciones Científicas, Spain
Prof. Michele Ceriotti, École Polytechnique Federale de Lausanne, Switzerland
Dr. Gregory Chasse, Queen Mary, University of London, United Kingdom
Prof. Emeritus Roger Cowley, University of Oxford, United Kingdom
Dr. Jesus Ivan da Silva Gonzalez, ISIS Facility, United Kingdom
Prof. John C. Dore, University of Kent, United Kingdom,
Dr. Christof Drechsel-Grau, Ruhr-Universität, Bochum, Germany
Prof. Felix Fernandez-Alonso, ISIS Facility and University College London, United Kingdom
Dr. Nikitas Gidopoulos, Durham University, United Kingdom
Prof. Giuseppe Gorini, Università Milano - Bicocca, Italy
Ms. Gabriella Graziano, University College London, United Kingdom
Prof. Neville Greaves, Aberystwyth University, United Kingdom.
Dr. Maria Grazia Izzo, Chalmers University of Technology, Sweden
Dr. Martin Jones, ISIS Facility, United Kingdom
Dr. Philip King, ISIS Facility, United Kingdom
Dr. Winfried Kockelmann, ISIS Facility, United Kingdom
Dr. Maciej Krzystyniak, Nottingham Trent University and ISIS Facility, United Kingdom
Prof. Dan Major, Bar-Ilan University, Israel
Dr. Jerry Mayers, University of Houston, United States of America
Prof. Robert McGreevy, ISIS Facility, United Kingdom
Mr. Joshua More, University of Oxford, United Kingdom
Dr. Sanghamitra Mukhopadhyay, ISIS Facility, United Kingdom
Dr. Stewart Parker, ISIS Facility, United Kingdom
Ms. Alexandra Parmentier, Università di Roma - Tor Vergata, Italy
Dr. Keith Refson, STFC Scientific Computing Department, United Kingdom
Dr. Selena Richards, Nottingham Trent University and ISIS Facility, United Kingdom
Mr. Giovanni Romanelli, Università di Roma - Tor Vergata, Italy
Dr. Svemir Rudic, ISIS Facility, United Kingdom
Dr. Kate Ryan, University of Oxford, United Kingdom
Dr. Christoph Salzmann, University College London, United Kingdom
Dr. Andrew Seel, ISIS Facility, United Kingdom
Dr. Roberto Senesi, Università di Roma - Tor Vergata, Italy
Dr. Lukasz Walewski, Ruhr-Universität, Bochum, Germany
Dr. Thor Wikfeldt, University of Iceland and Nordita, Iceland & Sweden
Mr. Tim Wiles, Durham University, United Kingdom

Journal of Physics: Conference Series **571** (2014) 011001 doi:10.1088/1742-6596/571/1/011001

References

[1] www.isis.stfc.ac.uk/news-and-events/events/2014

[2] www.osti.gov/scitech/biblio/5858602 and www.osti.gov/scitech/biblio/5860709

[3] www.fisica.uniroma2.it/vesuvio/dins/html/eV_workshop240405.htm

[4] www.ornl.gov

[5] web129.its.me.cnr.it/workshop/

[6] Mayers J and Reiter G 2012 *Meas. Sci. Technol.* **23** 045902

[7] www.isis.stfc.ac.uk/instruments/vesuvio

[8] Granroth GG, Kolesnikov I, Sherline TE, Clancy JP, Ross KA, Ruff JPC, Gaulin BD and Nagler SE 2010 *J. Phys. Conf. Ser.* **251** 12058

[9] www.neutrons.ornl.gov/sequoia

[10] Rodriguez Palomino LA, Blostein JJ and Dawidowski J 2011 *Nucl. Instrum. Meth. A* **646** 142

[11] www.cab.cnea.gov.ar

Journal of Physics: Conference Series **571** (2014) 011001 doi:10.1088/1742-6596/571/1/011001

Table 1. Programme of the *VI Workshop in Electron Volt Neutron Spectroscopy.*

Monday 20th January	
13:00	Welcome and Particulars
13:05-13:15	**Spectroscopy with Epithermal Neutrons** Felix Fernandez-Alonso, ISIS
Session 1	*Nuclear Quantum Effects*
13:15-13:55	**Quantal Nuclear Motions in Condensed H-Bonded Systems** Carla Andreani, Rome
13:55-14:35	**The Quantum Mechanics of Nano-Confined Water: New Cooperative Effects Revealed with Neutron and X-Ray Compton Scattering** George Reiter, Houston
Session 2	*DINS Beyond the Proton*
14:55-15:35	**MANSE and Beyond** Matthew Krzystyniak, ISIS
15:35-15:55	**Origins of Mechanical Toughness in Bio-Cement During Setting** Gregory Chasse, Queen Mary University of London
15:55-16:15	**Direct Measurement of Competing Quantum Effects on the Kinetic Energy of Heavy Water Upon Melting** Giovanni Romanelli, Rome
Session 3	*Measurement and Instrumentation*
16:20-16:50	**Epithermal Neutron Detection at ISIS** Guiseppe Gorini, Milano-Bicocca
16:50-17:10	**VESUVIO Now/VESUVIO When?** Andrew Seel, ISIS
17:10-17:35	**Direct and Inverse Geometry Instruments for Proton Momentum Distribution Studies** Roberto Senesi, Rome
Tuesday 21st January	
Session 4	*Theoretical Horizons and Calculation*
09:00-09:40	**Inexpensive Evaluation of the Quantum Kinetic Energy and Particle Momentum Distribution by First-Principles Simulations** Michele Ceriotti, EPFL
09:40-10:05	**Collective Proton Tunnelling in Ordinary Ice** Christoph Dreschel-Grau, Bochum
10:05-10:30	**Simulation Studies of Hydroge-Bonded Ferroelectrics** Thor Wikfeldt, Reykjavik/Stockholm
10:50-11:15	**Interaction Induced Localisation of Protons: Temperature and Superfluid Helium Solvation Effects** Lukasz Walewski, Bochum
11:15-11:40	**Nuclear Quantum Effects in Enzyme Catalysis From Momentum Distributions** Dan Major, Bar-Ilan
13:30-15:30	**Open Discussion**

Journal of Physics: Conference Series **571** (2014) 011002 doi:10.1088/1742-6596/571/1/011002

Peer review statement

All papers published in this volume of *Journal of Physics: Conference Series* have been peer reviewed through processes administered by the proceedings Editors. Reviews were conducted by expert referees to the professional and scientific standards expected of a proceedings journal published by IOP Publishing.

Journal of Physics: Conference Series **571** (2014) 012001 doi:10.1088/1742-6596/571/1/012001

The Quantum Mechanics of Nano-Confined Water: New Cooperative Effects Revealed with Neutron and X-Ray Compton Scattering

G. F. Reiter[1], **Aniruddha Deb**[2]

1Physics Department, University of Houston, 4800 Calhoun Houston, Tx, 77204-5505
2Department of Chemistry, University of Michigan, Ann Arbor, MI 48109

greiter@uh.edu

Abstract. Neutron Compton scattering(NCS) measurements of the momentum distribution of light ions using the Vesuvio instrument at ISIS provide a sensitive local probe of the environment of those ions. NCS measurements of the proton momentum distribution in bulk water show only small deviations from the usual picture of water as a collection of molecules, with the protons covalently bonded to an oxygen and interacting weakly, primarily electrostatically, with nearby molecules. However, a series of measurements of the proton momentum distribution in carbon nanotubes, xerogel, and Nafion show that the proton delocalizes over distances of 0.2-0.3Å when water is confined on the scale of 20Å. This delocalization must be the result of changes in the Born-Oppenheimer surface for the protons, which would imply that there are large deviations in the electron distribution from that of a collection of weakly interacting molecules. This has been observed at Spring-8 using x-ray Compton scattering. The observed deviation in the valence electron momentum distribution from that of bulk water is more than an order of magnitude larger than the change observed in bulk water as the water is heated from just above melting to just below boiling. We conclude that the protons and electrons in nano-confined water are in a qualitatively different ground state from that of bulk water. Since the properties of this state persist at room temperature, and the confinement distance necessary to observe it is comparable to the distance between the elements of biological cells, this state presumably plays a role in the functioning of those cells.

1. Introduction

The first indication that the quantum state of the protons in nano-confined water was unusual were measurements of the Debye-Waller factor in water confined in single wall carbon nanotubes(SWNT) at 4K[1]. The Debye-Waller factor was 5 times what it is in bulk ice at the same temperature. This implies a large zero point delocalization of the protons of 0.2-.03Å. This was confirmed by a neutron Compton scattering measurement[2] of the proton momentum distribution, which revealed a dramatic narrowing of the momentum distribution at 4K, relative to that of bulk water[3], persisting up to about 230K, with the room temperature distribution being approximately that of the bulk. This was attributed to structural changes in the water in the nanotubes at low temperature. Attempts to reproduce the momentum distribution of the protons at 4K, making use of what was believed to be the structure there and a high quality flexible, polarizeable, empirical potential model, failed. The model produced only small deviations from the momentum distribution of bulk water, and was unable to account for the 30% reduction in the momentum width that had been observed[4].

The next unexpected result was obtained on water confined in double walled carbon nanotubes(DWNT)[5]. The momentum distribution at 4K broadened, rather than narrowed, varied continuously with temperature, and remained anomalous at room temperature. This is particularly extraordinary, as the nominal inner dimensions of the SWNT, 14Å, and the DWNT, 16Å, are so close. The fact that the anomalous behavior persisted at room temperature, led to a new interpretation of an earlier experiment on water confined in xerogel[6], and subsequent experiments on water confined in Nafion, the electrolyte for some commercial fuel cell[5]. It was now conjectured that the changes in the momentum distribution observed were not due to the specifics of the interaction of the water with the confining surfaces, but were the effect of the confinement itself[5], that confinement led to a ground state distinct from the weakly interacting molecular ground state, and that the properties of this state persisted to room temperature. That the ground state of the electrons, oxygen ions and protons was distinct from the usual molecular ground state was confirmed by an x-ray Compton scattering experiment at room temperature in Nafion[7]. The momentum distribution of the valence electrons was measured, and found to be qualitatively different from that of bulk water. The difference at zero momentum is more than an order of magnitude greater than the differences in bulk water due to disordering of the hydrogen bond network as the water is heated from just above melting to just below boiling at atmospheric pressure. We conclude from these measurements, and others, that the state of nano-confined water is not describable by the usual molecular model, and conjecture that it involves the coherent motion of the protons, with a coherence length on the order of 10Å. We review below the experimental evidence and the reasoning behind that conjecture.

2. Neutron Compton Scattering

We show below a comparison of the neutron Compton profile, as measured on the VESUVIO instrument, for water confined in SWNT and DWNT, at 170K. To a first approximation, the interior of the carbon nanotubes is a smooth cylinder, providing only confinement, with the interaction usually modeled as a 6-12 potential. The diameters, 14Å and 16Å respectively, are sufficiently different to lead to a structure predicted by classical simulations at this temperature to be a cylinder of "square ice" with a chain of either one(SWNT) or three(DWNT) water molecules down the center[8]. Despite the small difference in the diameter, and in the classical structure, there is a great difference in the response of the protons to confinement, manifest in the pronounced narrowing of the momentum distribution in the SWNT case, and broadening in the DWNT case, compared to bulk water. (The momentum distribution for bulk ice at this temperature is similar to that of bulk water) The Compton profiles in both cases correspond to a delocalization of the protons in double well potentials. The potentials can be extracted, if we assume a mean field approximation for the interaction of the protons with themselves, by fitting the data with a simple model of a double well along the bond, and an harmonic well perpendicular to the bond[2]

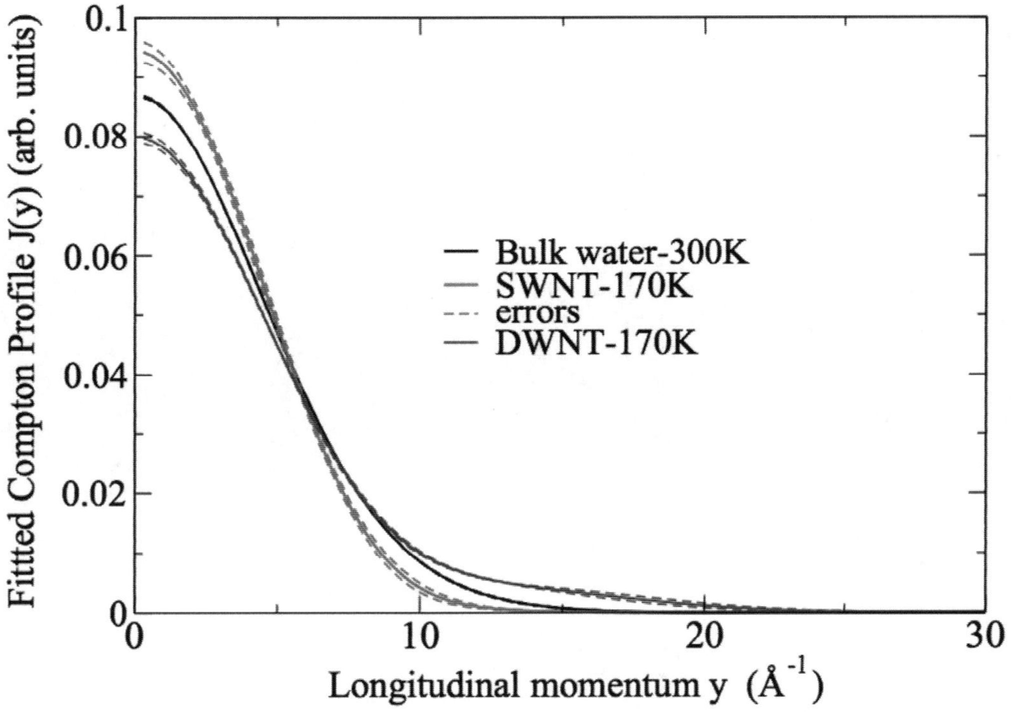

Figure 1. Comparison of the Compton profiles of bulk water, SWNT with diameter 14Å, and DWNT with diameter 16Å.

We show in Fig. 2 the variation with temperature of the effective potential obtained in this way for the DWNT. The increasing delocalization of the proton with temperature is evident from the figure. There is a structural change occurring at about 190K, as observed with the Debye Waller factor, similar to that observed in the SWNT. Particularly significant, from the point of view of the implications for biology, is that in this case, the double well potential is present at room temperature. The flat bottom potential observed at 4.2K is similar to that observed in the SWNT at the same temperature[2], although the overall width is different.

Although the effective potentials provide a convenient way of representing the momentum distribution data, it is clear that the delocalization of the proton is not an independent particle effect. As pointed out in references [1, 2], the vibrational band associated with the stretch mode is blue shifted in the SWNT, which cannot be accounted for with a single particle potential which softens to the extent observed. We take the pronounced difference shown in figure 1 to be additional indication of cooperative effects between the protons. If the momentum distribution was affected only by the ions in the immediate neighborhood of the proton, the momentum distribution would not be qualitatively different when the diameter of the nanotube changed by 13%. For this to happen, the change in the boundary conditions must be communicated throughout the material. While commensurability effects could account for this in the low temperature, ordered, structure, it would not be expected to do so in the room temperature phase, which is, if classical simulations are qualitatively correct, disordered in a way that is similar to bulk water.

The delocalization of the protons of water at room temperature has been observed in two other nano-confining systems, xerogel[9] and Nafion[5]. Variation of the pore size in the xerogel, allowed the conclusion that the effects of confinement were limited to scales on the order of 20Å.

Whatever the origin of the delocalization of the protons, the fact that they are delocalized implies that the many body Born-Oppenheimer potential that they are contained by is quite different from that of bulk water, where the proton is localized to about 0.1Å in the covalent bond with the oxygen. For this to be true, the spatial distribution of the valence electrons in the nano-confined water must be significantly different from that of bulk water. If that is the case, then the momentum distribution of the valence electrons must also be significantly different from that of bulk water.

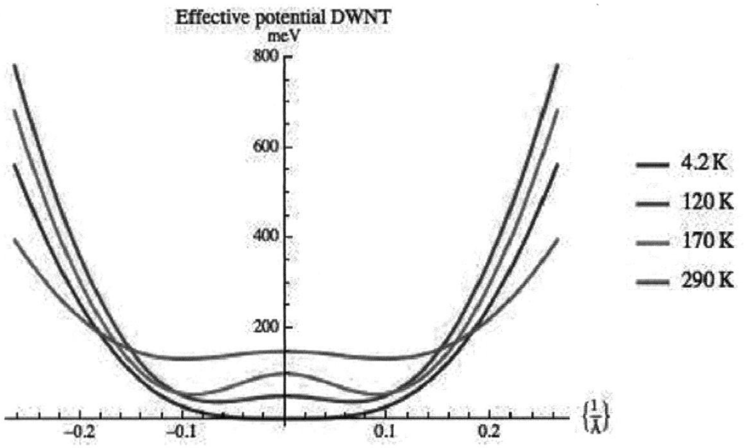

Figure 2. Temperature dependence of the effective potential for the protons in DWNT. The 120K, 170K, and 290K curves have been shifted up by 50, 100 and 150 meV respectively from the 4.2K curve.

3. X-Ray Compton Scattering

The momentum distribution of the electrons in the samples can be measured using X-ray Compton scattering. Measurements done at Spring-8 of water in Nafion at room temperature[7] show clearly that the changes in the valence electron distribution is large, far greater than can be accommodated within model for water that is based on a weakly interacting hydrogen bonded network of molecules. The technique measures the average momentum distribution of all the electrons in the system. Measurements on dry Nafion are used to subtract out the contribution from the electrons of the Nafion, and standard calculations for the momentum distribution for the core electrons of oxygen are used to obtain the contribution from the valence electrons. The resulting distribution was then subtracted from the measured distribution of bulk water to obtain the effects of confinement on the valence electrons. The results for two structurally different variants of Nafion are shown in figure 3.

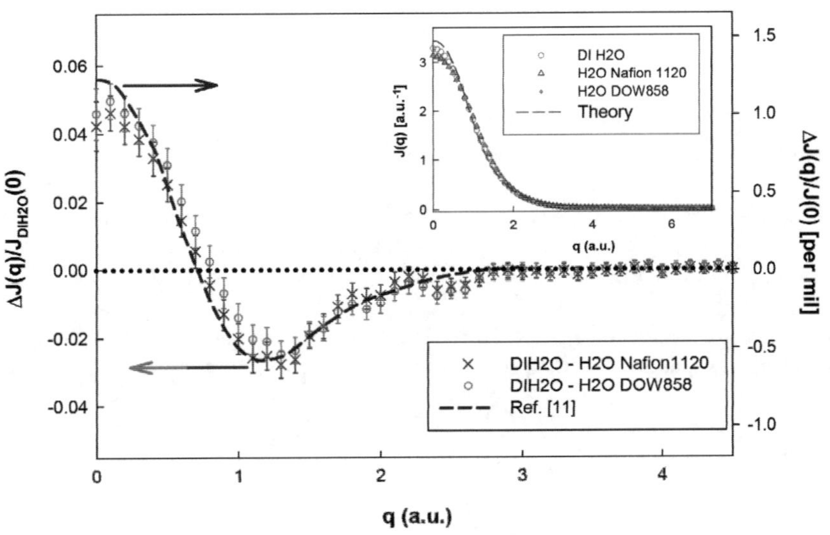

Figure 3. Comparison of the electron momentum distribution, measured with x-ray Compton scattering, of de-ionized bulk water (DIH2O) with that of water confined in two forms of Nafion. The main figure shows the difference between the Compton profile of the valence electrons in bulk water with that of the water confined in the Nafion samples. The fit is an empirical one. The shape of the fitted curve is that of the measured difference of the Compton profiles between deuterated and hydrogenated bulk water. The amplitude of the fitted curve is 46 times larger than that difference. The inset gives the un-subtracted measurements compared with a theoretical calculation for bulk water. The amplitude of the difference at q=0 is 17 times greater than the amplitude of the difference in bulk water between 5°C and 95°C

Calculations based on the molecular model, in which the variation with temperature of the electron momentum distribution are due to changes in the distribution of lengths and angles for the hydrogen bond[10], agree well with variation in the measured x-ray Compton scattering results on bulk water from just above freezing to just below boiling at room temperature. The variation in figure 3 between bulk and confined water at q=0 is 17 times the variation with temperature in bulk water at q=0, and cannot be explained within the molecular model. The phenomenological fit to the difference data shown in Fig. 3 is obtained by using the shape of the Compton profile from the difference between deuterated and hydrogenated bulk water[11], and fitting the amplitude to our data. The amplitude is 46 times larger, emphasizing again the large scale of the changes in the momentum distribution of the protons.

The disruption of the hydrogen bond network by confinement is far more profound than can be accounted for by varying the lengths and angles of the bonds. Together with the neutron measurements, the x-ray data demonstrates that the electron ion system in water is in a qualitatively different ground state from that of bulk water. We conjecture that is the correlated, cooperative motion of the protons, needed to explain the neutron data, that leads to the large changes in the electron distributions, and that these self consistently determine a Born-Oppenheimer potential that allows for the correlated motion.

4. Conclusions

There is a self consistent ground state of the electron proton-oxygen ion system, in which the protons are delocalized by 0.2-0.3Å, that is close in energy to the molecular ground state of bulk water. When the preferred structure of the hydrogen bonded network, a tetragonal arrangement of the water molecules, is sufficiently frustrated by confinement, this self-consistent delocalized state becomes the preferred ground state. The properties of this ground state are markedly different from that of bulk water. Measurements of the motion of free protons in the Nafion system, and the related system of reverse micelles[12], demonstrate that motion of the protons is so dramatically modified that it can no longer be called diffusive. Measurements of the decay of photo-excited marker ions in the water confined in Nafion[13] show the possibility of decay to states that would be in the band gap of bulk water.

In as much as the distance between the elements of biological cells is about 20Å, it seems likely that evolution has made use of the properties of this state in the functioning of our cells.

References

[1] Kolesnikov A I, Zanotti J-M, Loong C K, Thiyagarajan P, Moravsky A P, Loutfy R O, and Burnham C J, Phys. Rev. Letts. **93**, 35503 (2004)
[2] Reiter G, Burnham C, Platzman P M, Mayers J, Abdul-Redah T, Li J C, Loong C K, Kolesnikov A I, Phys.Rev. Letts. **97** 247801,(2006)
[3] Reiter G, Li J C, Mayers J, Abdul-Redah T, Platzman P, Braz. J. Phys., **34,** 142, (2004)
[4] Private communication, C J Burnham
[5] Reiter G F, Kolesnikov A I,Paddison S J et al. Phys. Rev. B Volume: **85**, 045403 (2012)
[6] Garbuio V, Andreani C, Imberti S, Pietropaolo A. and Reiter G F, J. Chem. Phys **127**, 154501 ((2007)
[7] Reiter G F, Deb A, Sakurai Y, Itou M, Krishnan V G, and Paddison S J, Phys. Rev. Letts. **111**, 036803 (2013)
[8] Ref. 1; Striolo A, Chialvo A A, Gubbins K E, Cummings P T, J. Chem. Phys. **122,** 234712 (2005)
[9] Garbuio V, Andreani C, Imberti S, Pietropaolo A, and Reiter G F, J. Chem. Phys **127**, 154501, ((2007)
[10] Hakala M, Nygard K, Manninen S, Huotari S, Buslaps T, Nillson A, Petersen L G M, Hamalainen K, J. Chem. Phys **125,** 084504 (2006)
[11] K Nygard, M Hakala, T Pylkkänen et al, J. Chem. Phys. **126**, 154508 (2007)
[12] Moilanen D E, Spry D B, and Fayer M D, Langmuir **24**, 3690 (2008)
[13] Tielrooij K J, Cox , M J, and Bakke H J, CPPC 10, **249** (2009)

Journal of Physics: Conference Series **571** (2014) 012002 doi:10.1088/1742-6596/571/1/012002

Mass-selective Neutron Spectroscopy Beyond the Proton

M. Krzystyniak[1,2], A.G. Seel [1], S.E. Richards[1,2], M.J Gutmann[1], and F. Fernandez-Alonso[1,3]

[1] ISIS Facility, Rutherford Appleton Laboratory, Chilton, Didcot, Oxfordshire OX11 0QX, United Kingdom

[2] School of Science and Technology, Nottingham Trent University, Clifton Campus, Nottingham NG11 8NS, United Kingdom

[3] Department of Physics and Astronomy, University College London, Gower Street, London WC1E 6BT, United Kingdom

Abstract. We discuss ongoing methodological developments underpinning the determination of nuclear-momentum distributions from mass-resolved neutron Compton data of lightweight materials. To this end, two systems are considered in detail, namely, lithium hydride (including its deuterated counterpart) and squaric acid, an organic antiferroelectric material containing hydrogen, carbon, and oxygen. Beyond the usual case of the proton, our approach enables direct access to detailed line shape information associated with the underlying nuclear-momentum distributions of both deuterium and lithium. For oxygen and carbon, mean kinetic energies can also be obtained directly from the neutron data, as demonstrated by a detailed analysis of mass-resolved data from squaric acid. From an instrumentation point of view, this work provides a suitable platform for a detailed assessment of existing capabilities and future developments in mass-selective neutron spectroscopy on the VESUVIO spectrometer at ISIS.

1. Introduction

Neutron Compton scattering (NCS) is an experimental technique made possible by ongoing developments in accelerator-based neutron sources [1]. As described in more detail in Refs. [1, 2], the use of electron-volt (eV) neutrons to measure atomic kinetic energies is predicated upon the fact that the energy (E) and momentum (Q) transferred in the scattering process are sufficiently high such that the so-called impulse approximation (IA) provides a convenient starting point for the analysis of experimental data. In the specific case of the VESUVIO neutron spectrometer at the ISIS Pulsed Neutron and Muon Source in the United Kingdom, typical E and Q lie in the range 1-100 eV and 30-200 Å^{-1}, respectively [3, 4]. In addition to VESUVIO, we also note recent developments in NCS instrumentation by Rodríguez Palomino *et al.* [5, 6] at the Bariloche Electron LINAC in Argentina.

In physical terms, the IA implies that the neutron scatters from a single atomic centre, with conservation of total kinetic energy and linear momentum. Consequently, NCS probes all motions associated with a given atom as given, for example, by the relevant atom-projected vibrational density of states (VDOS) [2]. Under these conditions, it follows that the measured scattering response in the so-called Deep-Inelastic-Neutron-Scattering (DINS) regime is related in a simple way to the nuclear-momentum distribution (NMD) of a given species. In this context, NCS experiments on single-crystal specimens have allowed the reconstruction of the proton wave

Journal of Physics: Conference Series **571** (2014) 012002 doi:10.1088/1742-6596/571/1/012002

function in much the same way as crystal structures may be obtained from diffraction data. Such a reconstruction is model independent and also allows a determination of the effective Born-Oppenheimer (BO) potential-energy surface of a given nuclide [2]. Conversely, comparison of experimental results from benchmark molecular and condensed-matter systems with *ab initio* computations can also provide stringent tests for state-of-the-art first-principles calculations, including recent developments in Path Integral Molecular Dynamics (PIMD) methodologies to describe nuclear-quantum effects in condensed matter [7].

An important consequence of the above relates to the intrinsic mass selectivity of NCS, a condition dictated by the requisite conservation of linear momentum during the scattering process, i.e., each atomic mass follows a distinct recoil line on the E-Q plane obeying the kinematic condition $E = \frac{\hbar^2 Q^2}{2M}$ where $\hbar$ is Planck's constant and M is the atomic mass. With sufficient spectral resolution (as that afforded by inverted-geometry neutron spectrometers like VESUVIO), the overall width of a given recoil line becomes sensitive to the chemical binding environment of a particular atom. Such mass and chemical selectivity (particularly for light nuclei) places NCS in a unique position relative to other scattering techniques. To date, the primary application of NCS has focused on the study of nuclear quantum effects associated with proton motions. Systems investigated include fundamental fluids and solids [8, 9, 10, 11, 12, 13, 14, 15, 16], metal hydrides [17, 18, 19, 20, 21], glasses and amorphous materials [22], hydrogen-bonded systems [11, 8, 12], proton conductors [23], or hydrogen-storage nanostructured materials [24]. Beyond the proton, recent instrument developments offer the exciting prospects of extending the realm of applicability of NCS to MAss-selective Neutron SpEctroscopic (MANSE) studies of heavier nuclei including lithium, boron, carbon, oxygen, or fluorine [8, 9, 17, 18].

MANSE studies have been reported recently on bulk and nanoencapsulated NaH [25], ^{7}LiF [26], lithium hydride (LiH) and its deuterated counterpart (LiD) [17, 18], and heavy water and ice [27]. In spite of the relatively low neutron-scattering cross sections of elements like Li, these studies have demonstrated sufficient sensitivity to the underlying NMD for this technologically relevant nuclide via a judicious choice of experimental conditions. Building upon these recent efforts, the present work seeks to provide the foundations of a general methodology for future MANSE studies. To this end, we assess the information content of NCS data measured on VESUVIO and extend existing data-analysis protocols for the treatment of MANSE data, always with a view to a detailed comparison with the predictions of first-principles calculations. Two systems are considered in detail, namely: LiH and its deuterated cousin LiD; and squaric acid (SA), an organic antiferroelectric material containing hydrogen, carbon, and oxygen.

2. Methodology

2.1. Theoretical Considerations

We recall the basic features of NCS, and the reader is referred to Refs. [1, 2] for further details. Within the IA introduced earlier, the dynamic structure factor $S(Q, E)$ for a given atom of mass M reduces to a single peak centered at the recoil energy $E_r = \hbar^2 Q^2/2M$ [28, 29, 30, 31, 32, 33, 34, 2]. Consequently, $S(Q, E) = (M/\hbar^2 Q)J(y)$, where y is the initial radial momentum p of the nucleus projected onto the scattering vector Q. $J(y)$ is the so called NCS profile [29, 28] and formally corresponds to the longitudinal nuclear momentum distribution (NMD) of the target nucleus along y. Mathematically [28, 35],

$$y = p \cdot \hat{Q} = \frac{M}{\hbar^2 Q}(E - E_r) = \frac{M}{\hbar^2 Q}\left(E - \frac{\hbar^2 Q^2}{2M}\right) \tag{1}$$

where $\hat{Q}$ is a unit vector along Q. Hereinafter, Q will be given in units of Å^{-1}, E in meV, and M in amu. With this choice of units, Plank's constant is given by $\hbar = 2.0446$ (meV amu)$^{1/2}$ Å.

In what follows, systems where $J(y)$ depends only on the magnitude of y but not on $\hat{Q}$ will be considered in the IA limit. However, in this class of systems we have to consider two cases: (i) systems where departures from the approximation of a harmonically bound isotropic nucleus are observed, either through anharmonicity or anisotropy of the local effective BO potential, and (ii) harmonically bound isotropic systems. In the second case, a spherically averaged three-dimensional longitudinal NMD, $J_{IA}(y)$, can be conveniently expressed via recourse to a Gram-Charlier expansion of the form [2, 36]

$$J_{IA}(y) = \frac{\exp(\frac{-y^2}{2\sigma^2})}{(\sigma\sqrt{2\pi})} \sum_n \frac{c_{2n}}{2^{2n} n!} H_{2n}(\frac{y}{\sigma\sqrt{2}}). \tag{2}$$

where H_{2n} denotes a Hermite polynomial and the standard deviation (STD) of a given NMD is denoted by σ. In Eq. (2), the term with $n = 0$ is set to unity due to the requisite normalization of the NMD. Moreover, with c_2 set to zero, the kinetic energy E_k of a nucleus of mass M is related to the STD of its NMD, σ, via $E_k = 3\hbar^2\sigma^2/2M$. This relationship is independent of the precise values of the remaining coefficients c_{2n}. Therefore, the fitting of $J_{IA}(y)$ to the Gram-Charlier expansion amounts to varying both the width of the momentum distribution σ and the coefficients c_{2n} with $n \geq 2$.

For a harmonically bound isotropic system where $J(y)$ depends only on the magnitude of y, the NCS profile in the IA limit can be written as a normalized Gaussian [34, 33, 2, 37, 30] which can be expressed as a special case of the Gram-Charlier expansion with coefficients $c_{2n} = 0$ for $n > 0$ in Eq. (2).

For finite values of Q, corrections to the IA are known as 'Final-State Effects' (FSEs) [34, 2]. To account for FSEs, the method of Sears [28] is routinely incorporated into standard NCS data treatments [34] by expressing the measured NCS profile $J(y)$ as a series of the form

$$J(y) = J_{IA}(y) + J_{FSE}(y) = J_{IA}(y) - \frac{M\langle\nabla^2 V\rangle}{36\hbar^2 q} \frac{d^3}{dy^3} J_{IA}(y) + ... \tag{3}$$

where $J_{IA}(y)$ is the IA result given by Eq. (2). $\langle\nabla^2 V\rangle$ is the mean value of the Laplacian of the potential energy of the nucleus expressed in meV Å^{-2} (cf. Refs. [34, 38]).

The STD of a given NMD, σ, as well as its associated Laplacian, $\langle\nabla^2 V\rangle$, may be calculated from the underlying VDOS under the assumption that the NMD for a given nucleus in a crystal along the $\widehat{Q}$ direction assumes a purely multivariate Gaussian functional form. This procedure is commonly referred to as the 'Gaussian' approximation (GA) [39, 2, 8, 17]. Similarly, the spherically averaged value of σ for nucleus n, σ_n, can be obtained using the appropriate atom-projected VDOS (PVDOS), $G_n(\omega)$, from [8]

$$\sigma_n^2 = \frac{M_n}{\hbar^2} \int G_n(\omega) \frac{\omega}{2} \coth\left(\frac{\omega}{2k_B T}\right) d\omega, \tag{4}$$

$G_n(\omega)$ for a harmonic solid can also be used to estimate the magnitude of FSEs [17] since [40]

$$\langle\nabla^2 V_n\rangle = \frac{3M}{\hbar^2} \int \omega^2 G_n(\omega) d\omega. \tag{5}$$

For the case of an isotropic three-dimensional harmonic oscillator (3D-HO) we can write

$$\langle\nabla^2 V\rangle = \frac{12\hbar^2\sigma^4}{M}. \tag{6}$$

Rewriting Eq. (3) using a new variable, $x = \frac{y}{\sqrt{2\sigma^2}}$, and using the definition of the third-order Hermite polynomial, $H_3(x) = 8x^3 - 12x$, we arrive at

$$J_{FSE}(x) = -\frac{k}{q}H_3(x)J_{IA}(x) \tag{7}$$

Then, an explicit connection between the parameter k and σ can be established for the case of a harmonically bound nucleus

$$k = \sigma\frac{\sqrt{2}}{12}. \tag{8}$$

Both σ and k or, more generally, the mean Laplacian of the potential energy of an given atomic nucleus, can be used as fitting parameters in the analysis of NCS data and then compared with a prediction based on an *ab initio* calculation. Since the second moment of a given NMD as well as the magnitude of FSEs can be calculated directly from the underlying VDOS, these two parameters can (in principle) be disentangled by devising analytical criteria constraining NCS observables.

2.2. Experimental Protocol

All NCS measurements were performed on the latest incarnation of the VESUVIO spectrometer at the ISIS Pulsed Neutron and Muon Source, United Kingdom [3, 4, 41]. As an inverted-geometry neutron spectrometer, the sample is exposed to a polychromatic neutron beam characterized by an incident-energy spectrum $I(E_0)$. Incident neutrons having initial energy E_0 travel a distance L_0 from the pulsed source to the sample. After scattering at an angle θ, neutrons of final energy E_1 travel a distance L_1 to the detector as schematically shown in Fig. 1. To achieve energy selection, a thin foil is placed in front of the detectors. This foil absorbs neutrons over a narrow range of energies, thereby fixing the final energy of the detected neutrons. All NCS measurements described in this work used a gold (Au) foil with a resonance energy of E_1= 4.9 eV. With this experimental arrangement, data collection necessarily involves foil cycling in both forward and backscattering geometries in order to obtain TOF data with and without energy selection, followed by a subtraction of these two data sets to obtain NCS spectra. In forward scattering, the energy-resolution function is adequately described by the convolution of a Lorentzian of half-width-at-half-maximum (HWHM) 24 meV and a Gaussian with STD 73 meV [42]. In backscattering, the resolution function is nearly Gaussian with extended (small) wings, and can be accurately described as the convolution of a Gaussian (STD=74 meV) and a Lorentzian (HWHM=24 meV) [43] For further details of the VESUVIO spectrometer, see the contribution by Seel *et al.* in the same issue.

The treatment of experimental NCS data is routinely performed in the time-of-flight (TOF) domain for a set of detectors placed at a series of scattering angles θ. For a given mass M, instrument-calibration procedures establish a unique relation between the values of Q, E, and y on each TOF channel t and scattering angle θ [43, 34]. With these considerations in mind, the total number of neutrons detected in an NCS spectrometer for a given mass M in a TOF channel t is proportional to the Compton profile, $J[y_M(t)]$, convolved in y_M space with a (mass-dependent) instrumental resolution function $R[y_M(t)]$. Thus, for a total of N different masses present in the sample, the total count-rate at a fixed scattering angle θ, $C_\theta(t)$, can be written as the following sum [cf. [34], Eq. (2.24)]:

$$C_\theta(t) = A'[\frac{E_0 I(E_0)}{Q}]_t \times \sum_{n=1}^{N} I_n M_n J_n[y_n(t)] \otimes R_n[y_n(t)] \tag{9}$$

Journal of Physics: Conference Series **571** (2014) 012002 doi:10.1088/1742-6596/571/1/012002

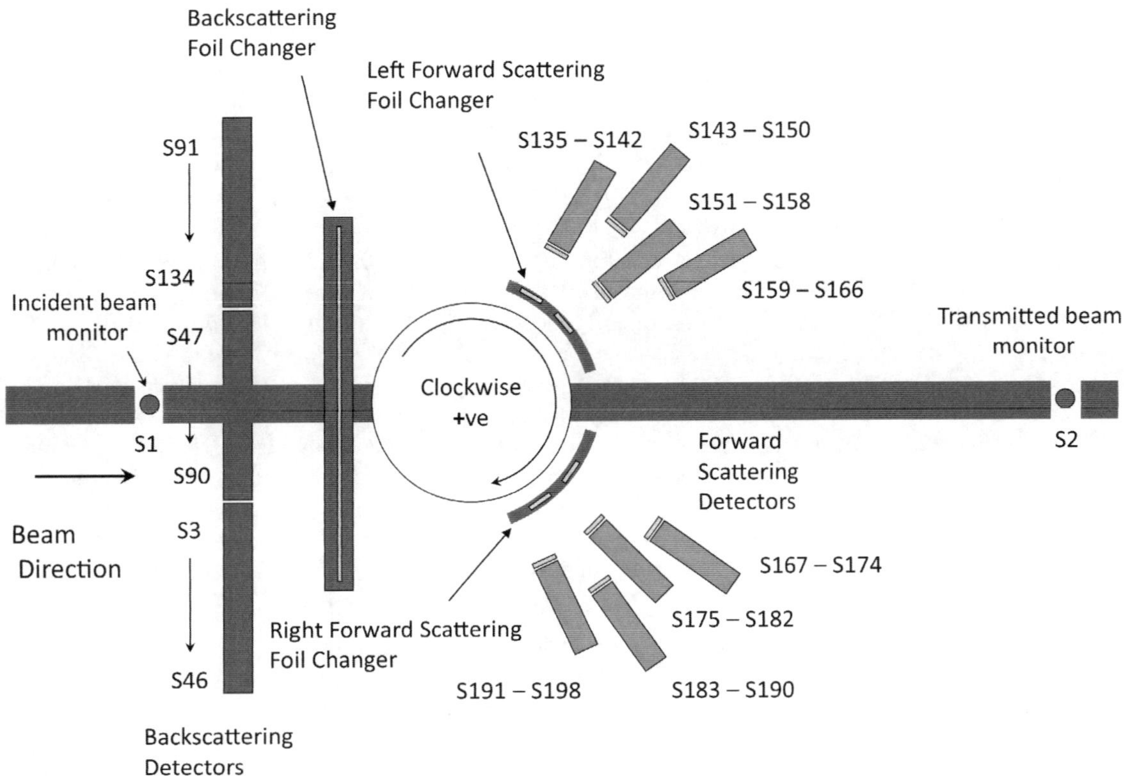

Figure 1. Schematic diagram of the VESUVIO spectrometer. For further details, see the main text and Seel *et al.* in the same issue.

where A' is a mass–independent experimental constant and the mass–independent factor $[\frac{E_0 I(E_0)}{q}]_t$ depends on the incident neutron spectrum, $I[E_0(t)]$, the initial neutron energy, $E_0(t)$, and the momentum transfer $Q(t)$, all explicit functions of t. In Eq. (9), the NMD for mass M, $J_M(y_M(t))$, is given by Eq. (3) with the expression for the NMD in the IA limit, $J_{IA}(y)$, given by Eq. (2). Integrated peak intensities I_M for a given scattering mass M are proportional to the scattering density $I_M = AN_M\sigma_M$, where $\sigma_M = 4\pi b_M^2$ is the total (bound) neutron-scattering cross section [44, 34].

For the NCS measurements on LiH and LiD, the specimens were enclosed in 6x6 cm^2 flat aluminum (Al) cells resulting in a total transmission of 85 and 90 % for LiH and LiD, respectively. In the case of SA, the sample was enclosed in a flat niobium (Nb) cell with 50-μm-thick windows leading to a total transmission of ca. 80 %. Use of thin Nb foils served to minimise contributions of the empty cell to the overall NCS response as well as spectral overlap with oxygen and carbon. All measurements reported herein were carried out over the range 4-300 K using a dedicated helium closed-cycle refrigerator.

3. Results and Discussion

3.1. Lithium Hydride and Deuteride

In recent works, Krzystyniak *et al.* [17, 18] have examined in detail the NCS response of both LiH and LiD, including a comparison of experimental NCS data with first-principles calculations and an assessment of the harmonic and impulse approximations over the kinematic range currently accessible on the VESUVIO spectrometer. As part of these efforts, a number of physical models for the underlying NMDs were compared with available experimental data, leading to the conclusion that for this particular system a purely Gaussian NMD and FSEs compliant with the 3D-HO model ($k = \sigma\frac{\sqrt{2}}{12}$) provide a satisfactory description of the NCS response for all nuclei in these materials. For LiH, it was found that σ_H=3.5±0.1 Å^{-1} and σ_{Li}=7.0±0.4 Å^{-1} in forward scattering, in excellent agreement with previous *ab initio* calculations (σ_H=3.57 Å^{-1} and σ_{Li}=7.14 Å^{-1}) [17]. For LiD, the same model gives σ_D=4.4±0.1 Å^{-1} and σ_{Li}=7.2±0.9 Å^{-1} in forward scattering. These values are within one STD from *ab initio* predictions, namely, σ_D=4.43 Å^{-1} and σ_{Li}=7.15 Å^{-1}. Although considerably less accurate, the widths of the Li momentum distributions in backscattering are also consistent with the above results, with values for σ_{Li} of 9.0±2.2 Å^{-1} and 7.0±0.8 Å^{-1} for LiH and LiD, respectively. Ref. [18] also explored the possibility of increasing the information content of NCS data by examining Cumulative Angle-Averaged Data (hereafter CAAD) in both forward and backscattering geometries. In this approach, this sum was directly compared with the corresponding sum of individual fits for a given detector at a given scattering angle θ using unit-area-normalised expressions for $C(\theta, t)$ given by Eq. (9), i.e., CAAD $= \sum_{i=1}^{N} \frac{C(\theta_i,t)}{\int C(\theta_i,t)dt}$. The summation range for this quantity corresponds to a selection of detectors after exclusion of outliers and faulty detectors. In practice, the selection process involves the removal of individual TOF spectra recorded by detectors which, when fitted with the expression given by Eq. (9), lead to reduced $\langle\chi^2\rangle$ values lower than 0.8 and higher than 1.2 [18]. This data-analysis procedure leads to a more transparent assessment of the performance of a given NMD model as well provides a convenient starting point for the development of 'detector-focusing' schemes aimed at increasing the effective count rate of TOF data relative to current data-analysis protocols.

As a natural extension of the above strategy, we consider the analysis of 'focused' NCS data directly obtained from a CAAD following the same criteria for the rejection of noisy and/or faulty detectors and outliers described in our previous work [18]. Figures 2 and 3 show LiH and LiD summed TOF data in forward and backscattering along with their respective fits. TOF data and associated error bars are shown in light grey, highlighting an order-of-magnitude improvement in signal-to-noise ratios (SNRs) relative to single-detector NCS data. These TOF data were pre-processed using the protocol described in detail in Ref. [18]. This procedure included the subtraction of multiple-scattering contributions, self-shielding and attenuation, and absorption. Contributions from γ backgrounds affecting the performance of the forward-scattering detector bank were removed using the standard VESUVIO data-reduction routines [45]. The data fits shown in these figures demonstrate effective mass separation of all nuclei in both forward and backscattering geometries. As visual aid, individual mass contributions have been colour-coded: blue for H and D in LiH and LiD, respectively; green for Li; and red for Al. The thick black lines in both figures describe the total TOF count rates, including sample container. The experimental data with associated error bars are shown as grey points.

Inspection of Figs. (2) and (3) demonstrates the adequacy of a physical model based on a Gaussian NMD with FSEs given by the 3D–HO model for both LiH and LiD in both forward and backscattering. The ratios of integrated signals associated with Li and Al, namely, $N_{Li}\sigma_{Li}/N_{Al}\sigma_{Al}$, read 0.8 and 2.8 for LiH and LiD, respectively. These ratios remain equal for forward and backscattering, further confirming the robustness of our data-analysis protocols. We note that, at present, CAAD fitting curves are generated assuming no TOF-dependent

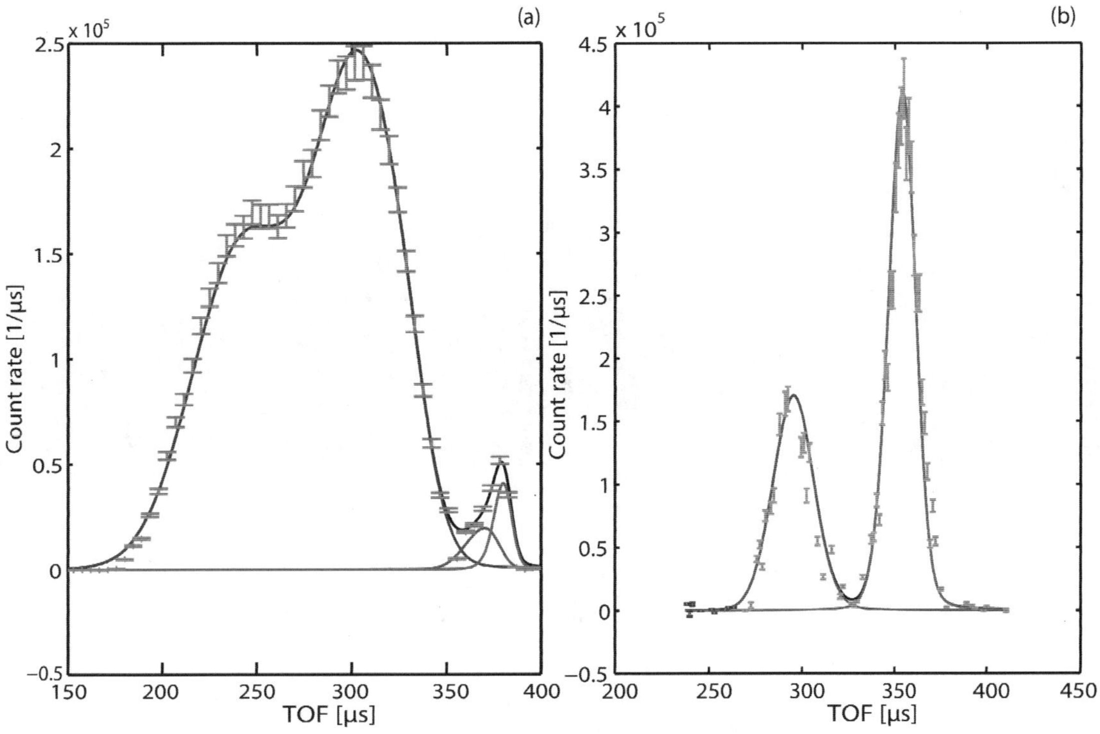

Figure 2. LiH CAADs and associated fits for (a) forward and (b) backscattering geometries. For further details, see the main text.

backgrounds, which typically show variations across detectors. With single-detector data, this effect is routinely accounted for by fitting TOF data with a third-order polynomial [45]. In the case of a cumulative fitting method like CAAD, the data measured at a given detector are first integrated in the TOF domain using numerical integration, then normalised by its TOF integral, and finally summed over the entire detector bank. In this procedure, information about the shape of each individual TOF-dependent background is then lost in the data-normalisation process. Consequently, the final CAAD signal contains a TOF-dependent background which cannot be traced back to specific background signals measured at a given detector. One obvious way to overcome this difficulty is to identify those detectors with negligible backgrounds. In some cases, however, this is not possible as the recorded TOF data, both in forward and backscattering, suffer from the presence of a TOF-dependent background in the region which physically corresponds to one or more recoil peaks. Such is the case of LiD, where we find that the D-recoil peak shape is affected by the presence of a strong background signal. Integration of the LiD TOF data consisting of D, Li, and Al recoil peaks superimposed on this background leads to an excess of the LiD CAAD signal compared to the LiD CAAD fitting curve around the TOF position of the D-recoil peak. This situation is shown in Fig. (3)a for forward scattering LiD TOF data. The discrepancy between the LiD CAAD signal and the CAAD fitting curve is even more pronounced in the case of LiD TOF backscattering data where the D-recoil peak is significantly distorted by the background. However, backscattering spectra showing no overlap between the D- and Li-recoil peaks can always be selected for the CAAD analysis, as shown in Fig. (3)b. Given

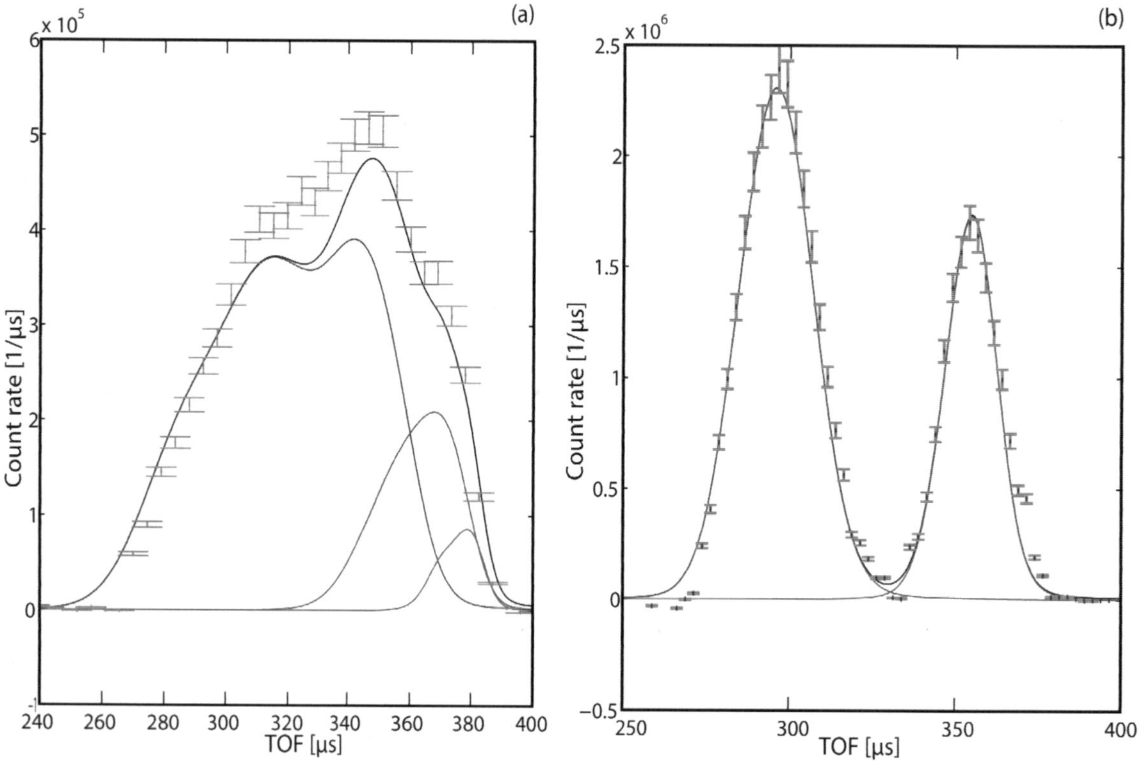

Figure 3. LiD CAADs and associated fits for (a) forward and (b) backscattering geometries. For further details, see the main text.

these limitations, only the width of the Li-recoil peak in LiD backscattering data was considered for comparison with previous studies. Overall, the NMD parameters obtained using the CAAD compare very well with those previously reported using other approaches [17, 18]. For LiH, we find σ_H=3.45±0.03 Å^{-1} and σ_{Li}=7.09±0.95 Å^{-1} in forward scattering, and σ_{Li}=7.02±0.25 Å^{-1} in backscattering. For LiD, σ_D=4.09±0.10 Å^{-1} and σ_{Li}=8.4±0.5 Å^{-1} in forward scattering, and σ_{Li}=6.4±0.4 Å^{-1} in backscattering. On the whole, the analysis of NCS data using a CAAD provides improved estimates of NMD parameters for Li in both forward and backscattering geometries relative to data-analysis protocols relying on far more sparse data associated with individual detectors. Forward and backscattering data for D in LiD still pose a number of challenges in either case owing to the presence of TOF-dependent experimental backgrounds. Work is in progress to improve the CAAD method to account for the presence of backgrounds in TOF data. Notwithstanding these limitations, forward-scattering data can still be used to provide reliable estimates for the underlying NMD and these results can still be used to check the consistency of the NCS data across detector banks. The essentially harmonic and highly isotropic character of the underlying VDOS in both LiH and LiD is ascertained by the analysis of NCS data presented in this Section. Extension of this work to highly anisotropic or anharmonic systems represents a natural extension of the above work in order to assess the loss of sensitivity to higher-order terms describing the underlying NMD, as NCS data are progressively averaged over scattering angle to generate a CAAD.

Journal of Physics: Conference Series **571** (2014) 012002 doi:10.1088/1742-6596/571/1/012002

3.2. Hydrogen, Carbon, and Oxygen in Squaric Acid

Encouraged by the results presented above on LiH and LiD, we have applied the same methodology to solid SA, a carbon-based antiferroelectric material containing H, C, and O (molecular formula $C_4O_4H_2$). In the solid, individual SA molecules are arranged in layers perpendicular to the *b*-axis, with strong hydrogen bonds within the $a-c$ plane. Below the onset of paraelectricity at 373 K, the four-membered carbon ring associated with a single molecular unit is distinctly asymmetric and hydroxyl (OH) groups across adjacent layers are arranged in an antiparallel fashion, giving rise to the observed antiferroelectricity. In parallel with the NCS measurements reported herein, we have performed detailed single-crystal neutron diffraction measurements on SA as a function of temperature using the SXD instrument at ISIS [46]. The temperature dependence of lattice constants and hydrogen-bond parameters shows an increasing decoupling of the crystal layers with temperature (expansion of the *b*-axis), and a concomitant weakening of the hydrogen bonds across individual SA units on the $a-c$ plane, a trend which starts well below room temperature. Given the above observations, SA constitutes a suitable molecular system to explore the sensitivity of NCS data to the anisotropic character of strong hydrogen bonds, as well as possible (and hitherto unexplored) changes in the nuclear dynamics of the heavier C and O species as a function of temperature.

The total NCS response of SA has been described as a angular sum of unit-area-normalised spectra accounting for the total count rate in TOF, i.e., CAAD $= \sum_{i=1}^{N} \frac{C(\theta_i,t)}{\int C(\theta_i,t)dt}$. Underlying NMDs for all nuclei were assumed to be Gaussian and the detector sum to run over all forward and backscattering detectors. FSEs were constrained within the 3D-HO model as shown by Eq. (8). In principle, the quality of the data allows for a determination of NMD parameters for the Nb cell used as sample container, although we note that these values are heavily dominated by the present resolution function of the instrument. Thus, Nb NCS widths at 18 K and at room temperature or above were constrained using tabulated Debye values corresponding to the low- and high-temperature limits [47]. These two limits give NMD STDs for Nb of 14 and 24.4 Å^{-1} at 18 K and 300 K, respectively. Additionally, CAAD fits in forward and backscattering were constrained to obey the expected 1:1 stoichiometry of C and O in SA. In order to check the feasibility of our MANSE approach in forward scattering for O- and C-containing systems, no constraints were imposed on the NMD STDs of these two nuclei, i.e., their numerical values were left free in the fitting sessions of both back- an forward-scattering data. Data reduction protocols including the removal of noisy and/or faulty detectors and outliers were identical to those employed for LiH and LiD in the previous Section.

MANSE spectra of SA along with their fits are shown in Figs. (4) and (5) for base (18 K) and room temperature (300 K), respectively. The experimental data with associated error bars are shown as grey points. The total fitted NCS response is shown in black, and the coloured lines represent contributions from individual nuclei: (i) blue for H; (ii) green for C; (iii) red for O; and violet for the Nb cell. At both temperatures, the fits provide an excellent description of the experimental NCS data, clearly showing the effective discrimination of all masses when both forward and backscattering data are analysed synchronously. Forward-scattering data can effectively discriminate between the NCS of the proton relative to other masses. In backscattering, the NCS response from C, O, and Nb can be separated into three distinct contributions.

To the best of our knowledge, these data represent the first instance where MANSE has been used with success to unambiguously separate up four different nuclei in a condensed matter system. Moreover, we note that a judicious choice of sample-environment conditions leads to a minor contribution from the container to the overall NCS response, a situation to be contrasted with the LiH and LiD data presented earlier. On the basis of these results, we anticipate that further technical developments on this front will be necessary for the development of MANSE into a fully fledged technique for the study of complex materials using NCS.

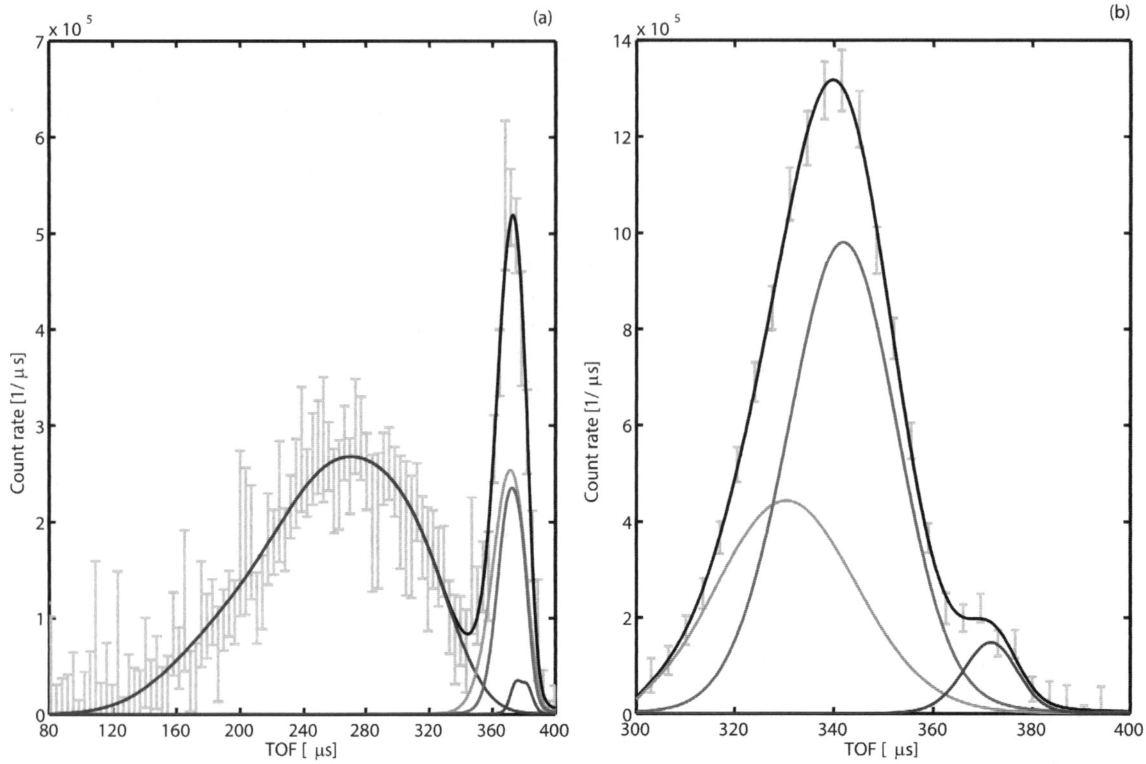

Figure 4. CAADs and associated fits for SA in (a) forward and (b) backscattering geometries at a temperature of T=18 K. For further details, see the main text.

From the fits of CAAD data in forward-scattering shown in Figs. (4) and (5), we obtain σ_H=4.6±0.2 Å^{-1} and 4.7±0.2 Å^{-1} at T=18 K and T=300 K, respectively. In comparison with LiH, these values represent a ca. 30 % increase in NCS widths, illustrating the sensitivity of NCS to an increase in 'chemical temperature' (binding strength) of the proton in SA relative to LiH in spite of its predominantly two-dimensional nature. Moreover, our results on SA demonstrate that the CAAD analysis followed in this work provides a convenient platform for MANSE studies of masses $M > 10$ amu. NMD STDs for C and O in SA at T = 18 K amount to σ_C=12±2.7(12±0.3) Å^{-1} and σ_O=12±2(11.4±0.3) Å^{-1} in forward(backscattering). At T = 300 K, σ_C=14±2(14±0.3) Å^{-1} and σ_O=14±2(14±0.3) Å^{-1} in forward(backscattering). Work is in progress to provide a basis for comparison of the above experimental values with first principles DFT results. As expected on the basis of previous considerations, the higher resolution of the backscattering detector banks leads to improved estimates of NMD STDs. Thus, a sensible data analysis strategy for MANSE data given current experimental capabilities on VESUVIO involves the use of forward-scattering CAADs for the analysis of proton-recoil data, and backscattering data for heavier nuclei, particularly those with masses greater than 10 amu. The contribution of Seel *et al.* in the same issue discusses a number of ways to circumvent these limitations in the future.

4. Outlook

The results presented herein provide a suitable platform for ongoing developments in MANSE using NCS techniques. We have shown that our data-analysis strategy enables the analysis of

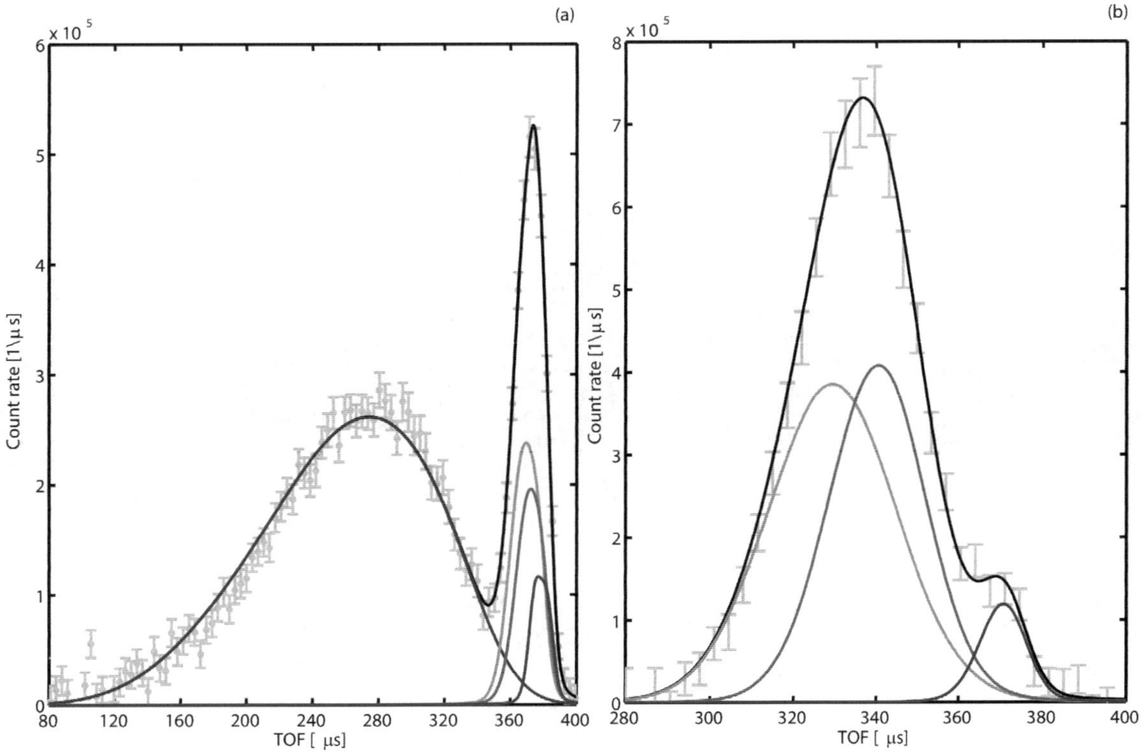

Figure 5. CAADs and associated fits for SA in (a) forward and (b) backscattering geometries at a temperature of T=300 K. For further details, see the main text.

MANSE data from composite materials containing H, D, Li, C, and O. Unlike neutron-scattering techniques using thermal and cold neutrons, neutron absorption by naturally occurring Li (or B) is substantially reduced in the epithermal (eV) region. Further work is needed to establish absolute sensitivity levels for these technologically relevant elements using MANSE, as well as a detailed assessment as to whether there is a real need for (typically costly) isotope enrichment protocols.

In the above context, we have also reported for the first time the analysis of MANSE data involving four distinct nuclei over the atomic-mass range 1-100 amu. MANSE data are also amenable to direct comparison with first-principles predictions and, as such, we anticipate that this much-needed synergy between NCS experiments and computational materials modelling will become an increasingly important area of activity in the foreseeable future. Extension of our approach to heavier masses and more complex materials requires both advances in instrumentation, to increase count rates and mass resolution, as well as an increasingly sophisticated and robust data-analysis toolkit to improve model-selection criteria.

5. Acknowledgments

The authors wish to thank J. Bones, R. Down, C. Goodway, R. Haynes, and M. Kibble from the ISIS Experimental Operations Division for their expert advice and superb technical support during the course of our experiments. The authors gratefully acknowledge the UK Science and Technology Facilities Council for financial support under Contract No. PR1200068, the ISIS Facility for access to beamtime, and the e-Science Facility at the Rutherford Appleton Laboratory for the provision of computing resources on the SCARF cluster.

[1] Fernandez-Alonso F and Price D L (eds) 2013 *Neutron Scattering – Fundamentals* (New York: Academic Press)
[2] Andreani C, Colognesi D, Mayers J, Reiter G F and Senesi R 2005 *Adv. Phys.* **55** 377
[3] Mayers J and Reiter G 2012 *Meas. Sci. Technol.* **23** 045902
[4] http://www.isis.stfc.ac.uk/instruments/vesuvio/
[5] Rodríguez Palomino L A, Blostein J J and Dawidowski J 2013 *J. Instrum.* **8** P08016
[6] Rodríguez Palomino L A, Blostein J J and Dawidowski J 2011 *Nucl. Instrum. Meth. A* **646** 142
[7] Lin L, Morrone J A, Car R and Parrinello M 2010 *Phys. Rev. Lett.* **105** 110602
[8] Krzystyniak M 2010 *J. Chem. Phys.* **133** 144505
[9] Krzystyniak M and Abdul-Redah T 2010 *Phys. Rev. B* **82** 064301
[10] C-Dreismann C A, Abdul-Redah T and Krzystyniak M 2005 *Phys. Rev. B* **72** 054123
[11] Abdul-Redah T, Krzystyniak M and C-Dreismann C A 2005 *Phys. Rev. B* **72** 052202
[12] C-Dreismann C A and Krzystyniak M 2006 *Journal of Physics: Condensed Matter* **18** 4741
[13] C-Dreismann C A and Krzystyniak M 2007 *Phys. Rev. B* **75** 057102
[14] Krzystyniak M, C-Dreismann C A, Lerch M, Lalowicz Z T and Szymocha A 2007 *Journal of Chemical Physics* **126** 124501
[15] Krzystyniak M, Lalowicz Z T, Szymocha A, C-Dreismann C A and Lerch M 2008 *Z. Anorg. Allg. Chem.* **634** 2056
[16] Krzystyniak M, Lalowicz Z T, C-Dreismann C A and Lerch M 2009 *J. Phys.: Condens. Matter* **21** 075502
[17] Krzystyniak M and Fernandez-Alonso F 2011 *Phys. Rev. B* **83** 134305
[18] Krzystyniak M, Richards S, Seel A and Fernandez-Alonso F 2013 *Phys. Rev. B* **88** 184304
[19] Abdul-Redah T, Krzystyniak M and C-Dreismann C A 2005 *Journal of Alloys and Compounds* **404** 790
[20] Abdul-Redah T, Georgiev P, Ross D K, Krzystyniak M and C-Dreismann C A 2005 *Journal of Alloys and Compounds* **404** 787
[21] Abdul-Redah T, Georgiev P, Krzystyniak M, Ross D K and C-Dreismann C A 2006 *Physica B* **385** 57
[22] C-Dreismann C A, Abdul-Redah T, Krzystyniak M and Vos M 2005 *Decoherence, Entanglement and Information Protection in Complex Quantum Systems (NATO Series II: Mathematics, Physics and Chemistry* vol 189) ed Akulin V M, Sarfati A, G Kurizki G and Pellegrin S (San Diego: Springer)
[23] Krzystyniak M and C-Dreismann C A 2007 *J. Phys.: Condens. Matter* **19** 436321
[24] Krzystyniak M, Adams M A, Lovell A, Skipper N T, Bennington S M, Mayers J and Fernandez-Alonso F 2011 *Faraday Discussions* **151** 171
[25] Seel A G, Sartbaeva A, Mayers J, Ramirez-Cuesta A J and Edwards P P 2011 *J. Chem. Phys.* **134** 114511
[26] Seel A G, Ceriotti M, Edwards P P and Mayers J 2012 *J. Phys.: Condens. Matter* **24** 365401
[27] Romanelli G, Ceriotti M, Manolopoulos D E, Pantalei C and Senesi R 2013 *J. Phys. Chem. Lett.* **4** 3251
[28] Sears V F 1984 *Phys. Rev. B* **30** 44
[29] Watson G I 1996 *J. Phys.: Condens. Matter* **8** 5955
[30] Mayers J, Andreani C and Baciocco G 1989 *Phys. Rev. B* **39** 2022
[31] Mayers J 1990 *Phys. Rev. B* **41** 41
[32] Mayers J 1993 *Phys. Rev. Lett.* **71** 1553
[33] Andreani C, Colognesi D and Pace E 1999 *Phys. Rev. B* **60** 10008
[34] Mayers J and Abdul-Redah T 2004 *J. Phys.: Cond. Matter* **16** 4811
[35] West G B 1975 *Phys. Rep. C* **18** 263
[36] Flammini D, Pietropaolo A, Senesi R, Andreani C, McBride F and Hodgson A 2012 *J. Chem. Phys* **136** 024504
[37] Evans A C, Timms D N, Mayers J and Bennington S M 1996 *Phys. Rev. B* **53** 3023
[38] Fielding A L, Timms D N and Mayers J 1998 *Europhysics Letters* **44** 255
[39] Senesi R, Colognesi D, Pietropaolo A and Abdul-Redah T 2005 *Phys. Rev. B* **732** 054119
[40] Sears V F 1973 *Physical Review A* **7** 340
[41] Pietropaolo A and Senesi R 2011 *Physics Reports* **508** 45
[42] Pietropaolo A, Andreani C, Filabozzi A, Pace E and Senesi R 2007 *Nucl. Instrum. Meth. A* **570** 498
[43] Mayers J and Adams M A 2011 *Nucl. Instrum. Meth. A* **625** 47
[44] Lovesey S W 1984 *Theory of Neutron Scattering from Condensed Matter* (Oxford: Clarendon)
[45] Mayers J 2011 User guide to vesuvio data analysis programs for powders and liquids Tech. Rep. RAL-TR-2011-003 Rutherford Appleton Laboratory, Chilton, UK URL http://purl.org/net/epubs/manifestation/5989
[46] Gutmann M J 2014 in preparation

Journal of Physics: Conference Series **571** (2014) 012002 doi:10.1088/1742-6596/571/1/012002

[47] Ho C Y, Powell R W and Liley P E 1974 *Thermal Conductivity of the Elements: A Comprehensive Review* (*J. Phys. Chem. Ref. Data* vol 3) (Washington D. C.: American Chemical Society and American Institute of Physics)

Journal of Physics: Conference Series **571** (2014) 012003 doi:10.1088/1742-6596/571/1/012003

The Harmonic Picture of Nuclear Mean Kinetic Energies in Heavy Water

G. Romanelli[1], **F. Fernandez-Alonso**[2,3] **C. Andreani**[1,4]

[1] Università degli Studi di Roma Tor Vergata, Dipartimento di Fisica and Centro NAST, via della Ricerca Scientifica 1, Rome 00133, Italy

[2] ISIS Facility, Rutherford Appleton Laboratory, Chilton, Didcot, Oxfordshire OX11 0QX, United Kingdom

[3] Department of Physics and Astronomy, University College London, Gower Street, London WC1E 6BT, United Kingdom

[4] Consiglio Nazionale delle Ricerche, CNR-IPCF, Sezione di Messina, I, Italy

E-mail: giovanni.romanelli@uniroma2.it

Abstract. This paper presents a study of the total mean kinetic energy, $\langle E_K \rangle$, and of individual projections along a given molecular axis, $\langle E_K \rangle_\alpha$, for D and O nuclei in D_2O, derived using a harmonic model. Our theoretical approach assumes decoupling amongst translational, rotational and vibrational modes. Resulting values of these dynamical quantities are discussed in terms of the anisotropy of the quantum kinetic energy tensor, its relation to the local potential, and deviations from the hypothesis of harmonicity and mode decoupling. Results are compared with corresponding quantities obtained from Deep Inelastic Neutron Scattering experiments performed on liquid and solid D_2O, where the short-time dynamics and local environment of D and O atoms were probed. The present study confirms an overall picture where even small changes in the short-range environment of D and O nuclei have a strong influence on the quantum behaviour of heavy water.

1. Introduction

The microscopic dynamics of liquid water still continues to challenge and attract great interest from theorists and experimentalists due to the fundamental relevance of this system in the biological and physical sciences [1]. A number of properties and anomalies of this molecule are associated with the presence of hydrogen bonds (HBs), influencing the structure and network in bulk ice and water, as well as its interaction with the surrounding environment. The quantum dynamics of hydrogen participating in HB influences the HB structure network, due to the non-commutativity of nuclear position and momentum operators, through nuclear quantum effects (NQEs) [2, 3]. Deep Inelastic Neutron Scattering (DINS), also known as Neutron Compton Scattering (NCS), is a unique and quantitative tool particularly well suited to probe the quantum behaviour of atomic nuclei. DINS allows one to highlight NQEs associated with HB breaking and distortion, through the measurement of single-particle dynamical quantities, *i.e.*, nuclear momentum distributions, $n(p)$, and mean kinetic energies, $\langle E_K \rangle$ [4–6]. Most recent DINS studies have shown that the merits and strengths of the technique have expanded beyond measurements of $n(p)$ for light elements (hydrogen, deuterium, helium, lithium), as it is now possible to obtain information on heavier nuclides such as carbon, oxygen, fluorine or sodium [4, 7–14].

In recent years, several *ab initio* Path Integral Molecular Dynamics (PIMD) simulations [15–18] and DINS experiments supplemented by PIMD simulations [19–28] have been performed to study the structure and dynamics of liquid water over a wide range of temperatures and pressures to investigate how these are directly influenced by NQEs. Moreover, some phenomenological models [29, 30] assuming harmonicity and decoupling amongst translational, rotational and vibrational modes have been used to calculate values of the total mean kinetic energy $\langle E_K \rangle$, as well as individual contributions $\langle E_K \rangle_\alpha$ along the three principal axes of the quantum kinetic energy tensor of the anisotropic $n(p)$. These models have been applied to describe the single-particle H dynamics in water [31–34] and in H_2S [35] by recourse to inter- and intra-molecular frequencies obtained from optical spectroscopy and Inelastic Neutron Scattering (INS).

In the following, we introduce a harmonic model for a generic triatomic molecule and discuss differences with other models in the literature. This model is used to predict kinetic energies for the particular case of D and O atoms in heavy water. These predictions are discussed and compared against data from a previous DINS experiment [8].

2. Harmonic model

Let us consider a molecule (D_2O) with the $N = 3$ atoms in the $y - z$ plane. For each of its $3N - 6 = 3$ internal modes, one can define a symmetry-adapted coordinate system (S_1, S_2, S_3) as illustrated in Figure 1. To first order, each atom undergoes harmonic motions along the normal coordinates S_i $(i = 1, 2, 3)$. Quantum mechanically, these motions necessarily lead to the emergence of a zero point energy proportional to the vibrational frequency ω_i. The potential energy of the system is the result of three pair interactions with restoring forces proportional to the displacement from the mean nuclear positions. The proportionality constant for the two D-O pairs is given by

$$k_{OD} = \frac{m_D \lambda_3 (1 + \mu \cos^2 \alpha)}{(1 + \mu \sin \alpha \cos \alpha)^2}, \tag{1}$$

where $\omega_3 = \sqrt{\lambda_3}$ is the asymmetric stretch frequency, $\mu = 2m_D/m_O$ and m_D and m_O the nuclear masses. Likewise, the proportionality constant for the D-D pair is given by

$$k_{DD} = \frac{m_D \lambda_2 \left[k_{OD}(1 + \mu \cos^2 \alpha) - m_D \lambda_2\right]}{2\left[k_{OD}(1 + \mu) \cos^2 \alpha - m_D \lambda_2\right]}, \tag{2}$$

with $\omega_2 = \sqrt{\lambda_2}$ being the bending frequency. Equations (1) and (2) are derived from the secular equation $|B - \lambda A| = 0$, with A and B being the kinetic- and potential-energy matrices expressed in the symmetry coordinates $\{S_1, S_2, S_3\}$, $\lambda_i = \omega_i^2$ are the eigenvalues of the matrix and $|\ldots|$ indicates the value of its determinant. The symmetry of the modes allows to express the dynamical 3×3 matrix as a $(2 \times 2)_\mathcal{S} \otimes (1 \times 1)_\mathcal{A}$ block matrix. Since only two observables, $\omega_2 = \sqrt{\lambda_2}$ and $\omega_3 = \sqrt{\lambda_3}$, are needed in order to fix the free parameters of the model, k_{OD} and k_{DD}, it is possible to evaluate the third eigenvalue of the secular equation, λ_1, and consider the difference with the measured value of ω_1^2 as an estimate of the accuracy of the model. In the case of heavy water, the symmetric stretching frequency evaluated within the present model is 20% higher than the measured one [36]. If one makes use of more than two parameters in an attempt to improve the accuracy of the model, then no prediction can be made on known observables and control on the accuracy of the model is lost.

The asymmetric $(1 \times 1)_\mathcal{A}$ block matrix has an eigenvector $s_\mathcal{A} = S_3$ that in Cartesian coordinates corresponds to a displacement vector $\xi_\mathcal{A} = (y_D, z_D, y_{D'}, z_{D'}, y_O, z_O) = (a, b, a, -b, -\mu a, 0)$, with $a = \sin \alpha$ and $b = \cos \alpha$, with no dependence on the stretching

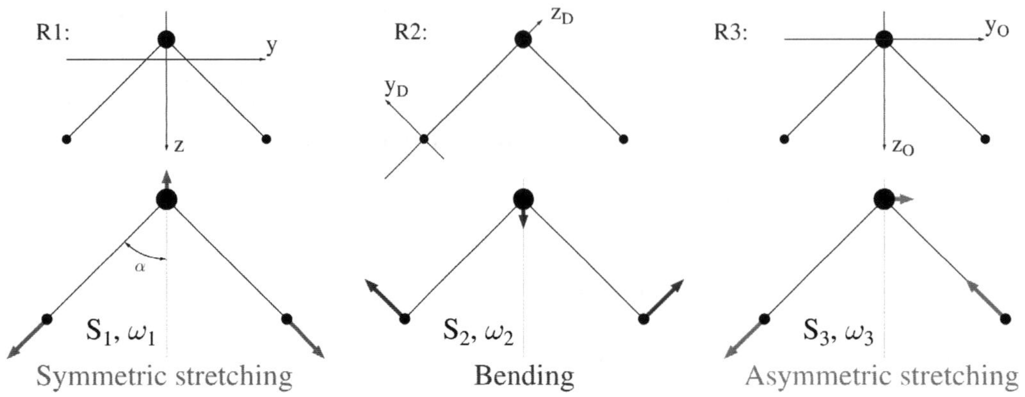

Figure 1. Definition of the symmetry coordinates for an isolated water molecule as used in the derivation of the coefficients f_i. The molecule lies in the $y-z$ plane and each axis may be oriented according to the reference frames R1, R2 and R3 (see text).

frequency. The symmetric $(2 \times 2)_{\mathcal{S}}$ block matrix has eigenvectors $s_{\mathcal{S}} = S_1 + v(\lambda)S_2$, with

$$v(\lambda) = \frac{\left(2k_{OD}\mu(1+\mu\cos^2\alpha) - 4k_{DD} - m_O\lambda\mu^2\right)\sin(2\alpha)}{8k_{DD}\cos^2\alpha + k_{OD}\mu^2\sin^2(2\alpha) - 2m_O\lambda\mu(1+\mu\sin^2\alpha)} \tag{3}$$

and with the limits $v(\lambda) \ll 1$ for $\lambda \to \lambda_1$ and $v(\lambda) \gg 1$ for $\lambda \to \lambda_2$. The corresponding displacement vector $\xi_{\mathcal{S}} = (c, d, -c, d, 0, -\mu d)$ in Cartesian coordinates is defined by $c = \sin(\alpha)+v(\lambda)\cos(\alpha)$ and $d = \cos\alpha - v(\lambda)\sin\alpha$. Each $i-th$ mode contributes to the kinetic energy of atom j with a fraction $f_{j,i} = \epsilon_{j,i}/\sum_j \epsilon_{j,i}$ of the mode energy, with $\epsilon_{j,i} = m_j(\dot{y}_{j,i}^2 + \dot{z}_{j,i}^2)/2$. The conditions imposed on the components of $\xi_{\mathcal{A}}$ and $\xi_{\mathcal{S}}$ reflect conservation of linear momentum in the two directions y and z only when these axes are defined as shown by reference frame R1 in Figure 1. However, the quantities $\epsilon_{j,i}$ depend only on the modulus of the vector (y_j, z_j) and then the energy fractions $f_{j,i}$ are invariant under axial rotation.

Following the previous discussion, one can numerically evaluate the parameters $f_{j,i}$ through Eqs. (1), (2) and (3) and the definition of $\xi_{\mathcal{S}}$ and $\xi_{\mathcal{A}}$. The calculation can be readily implemented in a computer routine and applied to any triatomic molecule. The sequential character of this procedure distinguishes the present model from others based on three independent force constants, the latter leading to a hard-to-solve system of coupled nonlinear equations.

Assuming an average frequency ω_τ corresponding to the translational modes, we can define $\tau_j = m_j/\sum_j m_j$. This parameter dictates the amount of energy that the $j-th$ atom draws from this particular vibration. Similar considerations apply to the rotational modes of frequency ω_ρ, whose contribution to the kinetic energy is defined by the ratio of moments of inertia $\rho_j = \bar{I}_j/\sum_j \bar{I}_j$. The quantity $\bar{I}$ corresponds to the spatial average of moments of inertia. With these definitions, the total mean kinetic energy of one atom (D or O) can be written as

$$\langle \mathcal{E}_K \rangle = 3\tau \langle \mathcal{E} \rangle_\tau + 3\rho \langle \mathcal{E} \rangle_\rho + \sum_{i=1}^{3} f_i \langle \mathcal{E} \rangle_i, \tag{4}$$

that is, as the sum of three translations, three rotations and three vibrations, each one with energy $\langle \mathcal{E} \rangle_s = \frac{\hbar\omega_s}{4}\coth\frac{\hbar\omega_s}{2k_BT}$, where $s = \tau, \rho, 1, 2, 3$. Let us now consider a reference frame

Journal of Physics: Conference Series **571** (2014) 012003 doi:10.1088/1742-6596/571/1/012003

		τ	ω_τ [meV]	ρ	ω_ρ [meV]	f_1	ω_1 [meV]	f_2	ω_2 [meV]	f_3	ω_3 [meV]	2α [degrees]
Ice	D	0.1	17.7	0.456	52.7	0.479	278	0.405	151	0.460	294	108.4
	O	0.8		0.088		0.042		0.190		0.079		
Water	D	0.1	15.1	0.457	46.9	0.499	296	0.410	150	0.458	310	106.0
	O	0.8		0.086		0.001		0.180		0.084		

Table 1. Optical frequencies and $\widehat{DOD}$ angles taken from Refs. [36–40], along with calculated values for τ, ρ, f_i using our model.

x,y,z (with the x axis perpendicular to the molecular plane) centred at the average position of each nucleus, representing the three principal axes of an anisotropic $n(\mathbf{p})$. The following interpretation of the DINS observables is based on the definition of a particular orientation of (y, z) for each atom. In the case of the D (Fig 1, R2), the z axis is taken in the direction of the stretching mode, *i.e.*, the mode that mostly contributes to the kinetic energy, the y axis in the direction of the bending mode and the rotation mode around the x axis. Rotations around y and z will contribute to the kinetic energy in the x direction. By considering an additional translational contribution along each direction, one can define for the D atom

$$\langle \mathcal{E}_K \rangle_x = \tau \langle \mathcal{E} \rangle_\tau + 2\rho \langle \mathcal{E} \rangle_\rho; \;\; \langle \mathcal{E}_K \rangle_y = \tau \langle \mathcal{E} \rangle_\tau + \rho \langle \mathcal{E} \rangle_\rho + f_2 \langle \mathcal{E} \rangle_2; \;\; \langle \mathcal{E}_K \rangle_z = \tau \langle \mathcal{E} \rangle_\tau + f_1 \langle \mathcal{E} \rangle_1 + f_3 \langle \mathcal{E} \rangle_3. \quad (5)$$

In the case of O, the most relevant contribution to the kinetic energy arises along the direction of the bending and the symmetric stretching modes (see Fig 1). Consequently, we take the z axis along this direction, while the y axis remains orthogonal and along the asymmetric stretching and the rotation around the x axis. For O one can then define

$$\langle \mathcal{E}_K \rangle_x = \tau \langle \mathcal{E} \rangle_\tau + \rho \langle \mathcal{E} \rangle_\rho; \quad \langle \mathcal{E}_K \rangle_y = \tau \langle \mathcal{E} \rangle_\tau + f_3 \langle \mathcal{E} \rangle_3; \quad \langle \mathcal{E}_K \rangle_z = \tau \langle \mathcal{E} \rangle_\tau + f_1 \langle \mathcal{E} \rangle_1 + f_2 \langle \mathcal{E} \rangle_2. \quad (6)$$

These contributions are calculated using frequencies from optical spectroscopy [36–40] as shown in Table 1. The corresponding $\langle \mathcal{E}_K \rangle$ and $\langle \mathcal{E}_K \rangle_\alpha$ are reported in Table 2. The values for the energy fractions τ, ρ, f_i have also been evaluated within the present model. These results are in good agreement with those reported in the literature [32, 33].

3. Results and Conclusion

We compare calculated values for $\langle \mathcal{E}_K \rangle$ and $\langle \mathcal{E}_K \rangle_\alpha$ with the results of DINS measurements on D_2O at $T = 274$ K (ice) and $T = 280$ K (liquid) [8], performed using the VESUVIO spectrometer at the ISIS Pulsed Neutron and Muon Source (Rutherford Appleton Laboratory, UK) [41, 42]. DINS measurements are conducted at high momentum and energy transfers ($\hbar q$ and $\hbar\omega$, respectively) thus allowing the scattering process to be described within the framework of the Impulse Approximation (IA) [43, 44]. The latter assumes that neutron scattering is incoherent and occurring within time scales much shorter than the typical relaxation times of collective excitations. Under these conditions, the struck particle recoils freely from the collision, with negligible inter-particle interactions in its final state (*i.e.*, the wave function of the particle in its final state is a plane wave). The incoherent dynamic structure factor $S(\mathbf{q}, \omega)$ is then expressed in terms of the single-particle momentum distribution $n(\mathbf{p})$

$$S_{IA}(\mathbf{q}, \omega) = \int n(\mathbf{p}) \delta \left(\omega - \omega_r - \frac{\mathbf{p} \cdot \hat{\mathbf{q}}}{M} \right) d\mathbf{p}, \quad (7)$$

with $\hbar\omega_r = \hbar^2 q^2 / 2M$ being the recoil energy and M the mass of the struck atom. One can then model the $n(\mathbf{p})$, whose variance is in turn related to the single-particle mean kinetic energy $\langle E_K \rangle$,

	$\langle E_K \rangle_x$ [meV]	$\langle \mathcal{E}_K \rangle_x$ [meV]	$\langle E_K \rangle_y$ [meV]	$\langle \mathcal{E}_K \rangle_y$ [meV]	$\langle E_K \rangle_z$ [meV]	$\langle \mathcal{E}_K \rangle_z$ [meV]	$\langle E_K \rangle$ [meV]	$\langle \mathcal{E}_K \rangle$ [meV]
D								
Liquid	18.8±1.1	15.1	38.6±2.5	22.9	54.2±2.4	73.6	112±2	112
Solid	22.5±1.8	15.9	37.4±2	23.8	48.1±3.4	68.3	108±2	108
O								
Liquid	16.0±2.3	10.4	21.0±0.6	16.1	24.1±2.1	16.4	61.1±3.1	43.4
Solid	16.2±2.3	11.0	20.1±1.6	15.4	24.2±1.4	19.7	60.4±4	46.1

Table 2. Comparison between experimental $\langle E_K \rangle$ and $\langle E_K \rangle_\alpha$ from Ref. [8] and our model using the parameters in Table 1.

through a spherical average of a multivariate Gaussian momentum distribution [8,17,35,45,46]

$$n(\mathbf{p}) = \frac{1}{\sqrt{8\pi^3 \sigma_x \sigma_y \sigma_z}} \exp\left(-\frac{p_x^2}{2\sigma_x} - \frac{p_y^2}{2\sigma_y} - \frac{p_z^2}{2\sigma_z} \right) \tag{8}$$

with x, y and z being the three molecular axes. This choice highlights the anisotropy of the local atomic potential. In particular, σ_α (with $\alpha = x, y, z$) represents the average nuclear momentum along a given axis. These definitions can be then used to define directional contributions to the kinetic energy $\langle E_K \rangle_\alpha = \sigma_\alpha^2/2M$ and a total mean kinetic energy $\langle E_K \rangle = \sum_\alpha \langle E_K \rangle_\alpha$. The resulting values for $\langle E_K \rangle$ and $\langle E_K \rangle_\alpha$ are reported in Table 2.

As far as D is concerned, our results show good agreement between $\langle \mathcal{E}_K \rangle$ and $\langle E_K \rangle$. In a recent study of liquid and solid light water [31], a similar agreement was found between the DINS value for $\langle E_K \rangle$ and $\langle \mathcal{E}_K \rangle$ derived from vibrational INS frequencies. Our model also reproduces quantitatively changes in directional kinetic energies associated with the liquid-to-solid phase transition, as observed by DINS. In this case, the decrease of the $z-$component in this transition is caused by an increase in HB strength in the solid phase – *i.e.*, stronger HBs lead to a red-shifted stretching frequency. Thus, both DINS experiment and our model agree on a small decrease by 5-6 meV of the kinetic energy along the z direction

The situation is markedly different for the component $\langle E_K \rangle_z$ and $\langle \mathcal{E}_K \rangle_z$. In this case, we find that the discrepancy between our model and DINS measurements is of approximately 20 meV (see Table 2). More generally, the observed differences between $\langle \mathcal{E}_K \rangle_\alpha$ and $\langle E_K \rangle_\alpha$ may be traced back to the assumption built into our model of decoupled motions when z is along the direction of the covalent bond (stretching mode). Moreover, intermolecular forces (neglected in our model) can lead to a redefinition of normal-mode coordinates. Interestingly, we note that much closer agreement with DINS may be achieved via a tilt of the z axis so as to maximize the contributions of the symmetric stretching and bending modes to $\langle \mathcal{E}_K \rangle_z$, resulting in a value for this energy of ca. 50 meV. Similar considerations would also apply to the y direction, with an associated kinetic energy of 40 meV. In summary while calculated and DINS values of $\langle \mathcal{E}_K \rangle$ and $\langle E_K \rangle$ yield information on the overall magnitude of NQEs, the directional components $\langle \mathcal{E}_K \rangle_\alpha$ and $\langle E_K \rangle_\alpha$ highlight the role of intermolecular interactions on single-particle momentum distributions.

From Table 2, one observes that for O atom the $\langle \mathcal{E}_K \rangle$ is underestimated of about 15 meV relative to $\langle E_K \rangle$ [8]. This finding may be ascribed to the long-range order of the HB network and to its effect on translational motions. It is entirely plausible that the present model underestimates these effects: for example, in Table 1 one can see that the contribution from translational modes dominates $\langle \mathcal{E}_K \rangle$.

In conclusion, comparison between our model and DINS shows that $\langle E_K \rangle$ for D in D_2O may be determined by combining translational, rotational and vibrational frequencies from optical spectroscopy and INS, in agreement with previous work on light water [31]. These results suggest that a combined use of optical spectroscopy, INS and DINS measurements can give simultaneous access to the magnitude of underlying NQEs in water, as well as both local and intermediate-range-scale properties.

The implementation of the present model as part of the VESUVIO data analysis suite may be also used as a predictive tool in experimental design. As such, it would pave the way for detailed exploration of directional parameters σ_α using a multivariate Gaussian line shape for $n(p)$. Further, this method could in principle be generalized to more complex materials.

Acknowledgement. This work was partially supported within CNR-STFC Agreement No. 06/20018 concerning collaboration in scientific research at the ISIS Pulsed Neutron and Muon Source (Rutherford Appleton Laboratory, UK).

References

[1] Kühne T D and Khaliullin R Z 2013 *Nat. Commun.* **4** 1450
[2] Morrone J A and Car R 2008 *Phys. Rev. Lett.* **101** 017801
[3] Li X Z, Walker B and Michaelides A 2011 *Proc. Natl. Acad. Sci. U.S.A.*
[4] Reiter G F, Senesi R and Mayers J 2010 *Phys. Rev. Lett.* **105** 148101
[5] Senesi R, Pietropaolo A, Bocedi A, Pagnotta S E and Bruni F 2007 *Phys. Rev. Lett.* **98** 138102
[6] Senesi R, Colognesi D, Pietropaolo A and Abdul-Redah T 2005 *Phys. Rev. B* **72** 054119
[7] Flammini D, Pietropaolo A, Senesi R, Andreani C, McBride F, Hodgson A, Adams M A, Lin L and Car R 2012 *J. Chem. Phys* **136** 024504
[8] Romanelli G, Ceriotti M, Manolopoulos D E, Pantalei C, Senesi R and Andreani C 0 *J. Phys. Chem. Lett.* **0** 3251–3256
[9] Krzystyniak M and Fernandez-Alonso F 2011 *Phys. Rev. B* **83** 134305
[10] Krzystyniak M and Abdul-Redah T 2010 *Phys. Rev. B* **82** 064301
[11] Krzystyniak M 2010 *J. Chem. Phys* **133** 144505
[12] Krzystyniak M, Adams M A, Lovell A, Skipper N T, Bennington S M, Mayers J and Fernandez-Alonso F 2011 *Farad. Discuss.* **151**(0) 171–197
[13] Seel A G, Ceriotti M, Edwards P P and Mayers J 2012 *J. Phys.: Condens. Matter* **24** J5401
[14] Krzystyniak M, Richards S, Seel A and Fernandez-Alonso F 2013 *Phys. Rev. B* **88** 184304
[15] Lee C, Yang W and Parr R G 1988 *Phys. Rev. B* **37** 785
[16] Becke A D 1988 *Phys. Rev. A* **38** 3098
[17] Goedecker S, Teter M and Hutter J 1996 *Phys. Rev. B* **54** 1703–1710
[18] VandeVondele J, Krack M, Mohamed F, Parrinello M, Chassaing T and Hutter J 2005 *Comp. Phys. Comm.* **167** 103–128
[19] Andreani C, Colognesi D, Mayers J, Reiter G F and Senesi R 2005 *Adv. Phys.* **54** 377–469
[20] Morrone J A, Lin L and Car R 2009 *jcp* **130** 204511
[21] Andreani C, Colognesi D, Degiorgi E and Ricci M A 2001 *J. Chem. Phys* **115** 11243–11248
[22] Pietropaolo A, Senesi R, Andreani C, Botti A, Ricci M A and Bruni F 2008 *Phys. Rev. Lett.* **100** 127802
[23] Pietropaolo A, Senesi R, Andreani C and Mayers J 2009 *Braz. J. Phys.* **39** 318–321
[24] Andreani C, Colognesi D, Pietropaolo A and Senesi R 2011 *Chem. Phys. Lett.* **518** 1 – 6 ISSN 0009-2614
[25] Colognesi D, Andreani C and Degiorgi E 2003 *J. Neutron Res.* **11** 123–143
[26] Ceriotti M, Bussi G and Parrinello M 2009 *Phys. Rev. Lett.* **103** 030603
[27] Lin L, Morrone J A, Car R and Parrinello M 2010 *Phys. Rev. Lett.* **105** 110602
[28] Pantalei C, Pietropaolo A, Senesi R, Imberti S, Andreani C, Mayers J, Burnham C and Reiter G 2008 *Phys. Rev. Lett.* **100** 177801
[29] Moreh R, Shahal O and Volterra V 1976 *Nucl. Phys. A* **262** 221–230
[30] Vos M, Weigold E and Moreh R 2013 *J. Chem. Phys* **138** 044307
[31] Andreani C, Romanelli G and Senesi R 2013 *J. Chem. Phys* **427** 106–110
[32] Moreh R and Nemirovsky D 2010 *J. Chem. Phys* **133** 084506
[33] Finkelstein Y and Moreh R 2014 *Chem. Phys.* **431** 58–63
[34] Senesi R, Flammini D, Kolesnikov A I, Murray É D, Galli G and Andreani C 2013 *J. Chem. Phys* **139** 074504
[35] Andreani C, Degiorgi E, Senesi R, Cilloco F, Colognesi D, Mayers J, Nardone M and Pace E 2001 *J. Chem. Phys* **114** 387–398
[36] Max J J and Chapados C 2002 *J. Chem. Phys* **116** 4626–4642
[37] Bertie J E and Whalley E 1964 *J. Chem. Phys* **40** 1637–1645

[38] Taylor M J and Whalley E 1964 *J. Chem. Phys* **40** 1660–1664
[39] Fanourgakis G S and Xantheas S S 2006 *J. Chem. Phys* **124** 174504
[40] Shawyer R E and Dean P 1972 *J. Phys. C* **5** 1028
[41] Senesi R, Andreani C, Bowden Z, Colognesi D, Degiorgi E, Fielding A L, Mayers J, Nardone M, Norris J, Praitano M, Rhodes N J, Stirling W G, Tomkinson J and Uden C 2000 *Phys. B* **276** 200–201
[42] Mayers J and Reiter G 2012 *Meas Sci Technol* **23** 045902
[43] West G B 1975 *Phys. Rep.* **18** 263–323
[44] Gunn J M F, Andreani C and Mayers J 1986 *J. Phys. C* **19** L835–L840
[45] Ceriotti M, Miceli G, Pietropaolo A, Colognesi D, Nale A, Catti M, Bernasconi M and Parrinello M 2010 *Phys. Rev. B* **82** 174306
[46] Lin L, Morrone J A, Car R and Parrinello M 2011 *Phys. Rev. B* **83** 220302

Journal of Physics: Conference Series **571** (2014) 012004 doi:10.1088/1742-6596/571/1/012004

Discussion: Nuclear Quantum Dynamics - Protons and Beyond

C. Andreani,[1,2] M. Ceriotti,[3] G. Chass,[4] C. Drechsel-Grau,[5] F. Fernandez-Alonso,[6,7] N. Greaves,[8,9] N. Gidopolous,[10] M. Krzystyniak,[6,11], A. Parmentier,[1] K. Refson,[6] G.F. Reiter,[12] G. Romanelli,[1] A. Seel,[6,13] L. Walewski,[5] T. Wiles.[10]

[1] Università degli Studi di Roma Tor Vegata, Dipartimento di Fisica and Centro NAST, via della Ricerca Scientifica 1, Rome 00133, Italy
[2] CNR-IPCF Sezione di Messina, Messina, Italy
[3] Laboratory of Computational Science and Modelling, NQ Institute of Materials, École Polytechnique Fédérale de Lausanne, CH-1015 Lausanne, Switzerland
[4] School of Biological and Chemical Science, Queen Mary University of London, London E14NS, United Kingdom
[5] Lehrstuhl für Theoretische Chemie, Ruhr-Universität Bochum, 44780 Bochum, Germany
[6] ISIS Facility, Rutherford Appleton Laboratory, Chilton, Didcot, Oxfordshire OX11 0QX, United Kingdom
[7] Department of Physics and Astronomy, University College London, Gower Street, London WC1E 6BT, United Kingdom
[8] Centre for Advanced Functional Materials and Devices, Institute of Mathematics and Physics, Aberystwyth University, Aberystwyth SY23 3BZ, United Kingdom
[9] Department of Materials Science and Metallurgy, University of Cambridge, Cambridge BB2 3QZ, United Kingdom
[10] Department of Physics, Durham University, South Road, Durham DH1 3LE, United Kingdom
[11] School of Science and Technology, Nottingham Trent University, Clifton Campus, Nottingham NG11 8NS, United Kingdom
[12] Physics Department, University of Houston, 4800 Calhoun Houston, Texas 77204-5505, United States of America
[13] Department of Chemistry, University of Oxford, South Park Road, Oxford OX1 3QR, United Kingdom

1. Quantal Nuclear Motions in Condensed H-bonded Systems (Carla Andreani, Rome, Italy)

Dr Refson opened the discussion of the lecture by Prof Andreani: Which are the next experimental prospects to include other phases of water ice?
Prof Andreani replied: Next 3-5 years will be critical. The temperature trend of ice is not well understood, yet so full of meaning. The convergence of theory, INS and DINS is needed to characterize anharmonicity. Three major investigation areas: ice, liquid water, supercooled water.

Prof Greaves communicated: Have you considered levitation techniques to reach supercooled

state for water? Below the melting point there are anomalous changes in compressibility and C_p which might be mirrored in proton energy changes. An interesting issue is the trends. These reveal the direction of a possible second critical point for water.
Prof Andreani communicated in response: I'm very interested in the possibility of using levitation methods. As to the second critical point, my group is experimentally trying to probe that region starting from amorphous phases of solid water.

Dr Krzystyniak queried: Are there any hints about anomalous mean kinetic energy from ^{17}O NMR?
Prof Andreani replied: Not to my knowledge

Dr Krzystyniak remarked: Could one measure O-projected VDOS on TOSCA?
Prof Andreani replied: There is no reliable way to do that due to no variation in Q on such a machine.

Mr Wiles queried: Is there a chance to make simultaneous measurements on VESUVIO?
Prof Andreani remarked: There is a chance to derive some vibrational information, but with low resolution. MARI has a better resolution, as well as SEQUOIA.

Prof Fernandez-Alonso communicated: Is there a minimum in the mean kinetic energy of liquid water at about 300 K? What is its origin?
Prof Andreani communicated in response: The origin is not clear yet. Data should be remeasured. The region which actually needs to be investigated best is the supercooled one, as well as the one around 280 K.

Mr Wiles queried: Do you use any other spectroscopic techniques on the ice sample in situ (e.g., diffraction)? This is important as the sample changes rapidly with time and is inhomogeneous in my understanding.
Prof Andreani replied: The sample is homogenous for the time scale of the measurements, and we do not use other spectroscopic techniques in situ.

2. The Quantum Mechanics of Nano-Confined Water: New Cooperative Effects Revealed with Neutron and X-Ray Compton Scattering (George Reiter, Houston, TX, USA)

Prof Gidopoulos opened the discussion of the lecture by Prof Reiter: How do you distinguish external from internal water in nanotubes?
Prof Reiter replied: The nanotubes come in bundles with a hexagonal lattice. A Bragg peak of this lattice is nearly extinguished by the form factor of the nanotubes. The form factor of the nanotubes depends on the neutron dielectric constant of the water. The resultant small peak from the cancellation of the Bragg peak by the form factor would shift in the opposite direction from that observed upon filling if the water was on the outside of the nanotube.

Prof Greaves communicated: Confinement reduces the melting point. Is there a link between confinement in pores of different dimensionality (such as nanotubes vs Nafion)?
Prof Reiter communicated in response: We have considered this issue. It appears like it is a scale-matter, in the sense that the 20Å correlation length dominates.

Journal of Physics: Conference Series **571** (2014) 012004 doi:10.1088/1742-6596/571/1/012004

Prof Greaves commented: Is shape not relevant above 20Å?
Prof Reiter replied: Exactly.

3. MANSE and Beyond (M. Krzystyniak, STFC, UK))

Dr Ceriotti opened the discussion of the lecture by Dr Krzystyniak: Are there indications of deviations from the Maxwell-Boltzmann distribution for Cs?
Dr Krzystyniak replied: At room temperature, Cs behaves pretty much as an ideal classical gas from the n(p) point of view. However all and by itself the measurement of the kinetic energy of a heavy nucleus directly from a resonance opens up a new way for electron-volt spectroscopy with neutrons For the first time, widths of momentum distributions of some heavy nuclei, measured directly from time of flight spectra recorded on VESUVIO, can be compared with *ab initio* calculation

Dr Refson communicated: If I understand correctly, you are using discrepancy between harmonic phonon calculation and experimental n(p) as a measure of the size of non-adiabatic effects. But anharmonicity also contributes to this. How large is the anharmonic contribution?
Dr Krzystyniak communicated in response: Actually both LiH and LiD are very harmonic. In a recent paper [*Phys. Rev. B*, **87**, 144302 (2013)], a formalism was described for studying temperature dependent anharmonic vibrational properties in periodic systems using first-principles calculations. Within this formalism, the BO energy surface is mapped using a principal axes approximation, and the resulting equations are solved using a Hartree mean-field approach and second-order perturbation theory. Following this, the vibrational anharmonic wave function is used to calculate expectation values of phonon-dependent quantities. The authors have tested the formalism on diamond, lithium hydride, and lithium deuteride, which exhibit quite small anharmonicities. However, while diamond is nearly harmonic, the small anharmonicity in lithium hydride arises from a cancellation of the contributions from single-phonon and two-phonon terms.

Dr Walewski queried: Is the momentum distribution corresponding to the total VDOS? Is it possible to measure the anisotropy of the momentum distribution in the DINS experiments?
Dr Krzystyniak replied: Yes, the momentum distribution corresponds to the VDOS, and the anisotropy can be measured.

Dr Seel remarked: Are there factors other than the thermal broadening that affect the width of the recoil, excluding resolution effects? Is there only a single decay process?
Dr Krzystyniak replied: The instrumental resolution contribution to a resonance peak measured on VESUVIO can be calculated analytically in the limit of the vanishing final flight path L1 and the thick sample limit. Comparison of the synthetic VESUVIO re- sponse including the resolution function calculated in this limit for the case of Cs resonances in CsHSO4 with experimental data on shows that any possible contributions from VESUVIO analyser foils (primary and secondary) is negligible. The Doppler broadening of resonances of caesium in CsHSO4 as observed on VESUVIO does not seem to be dependent on decay processes. Different resonances observed for different incident neutron energies show the same extent of Doppler broadening.

Journal of Physics: Conference Series **571** (2014) 012004 doi:10.1088/1742-6596/571/1/012004

4. Origins of Mechanical Toughness in Bio-Cement During Setting (Gregory Chass and Neville Greaves, QMUL,Aberystwyth University, UK

Prof Fernandez-Alonso opened the discussion of the lecture by Prof Chass and Prof Greaves: It is quite remarkable that the mechanical (macroscopic) properties can be linked to (atomic-scale) NCS data. Can you further comment on this?
Dr Chass replied: Mechanical toughness in composites derives from interfaces, and it is dominated by coordination.

Prof Fernandez-Alonso communicated: Is then the interface predominant in these materials, to be so clearly seen in the Compton data?
Dr Chass replied affirmatively.

Prof Andreani commented: What do you need for future experiments? Will D_2O help?
Prof Chass replied: D_2O would be useful. Also we need to explore different temperatures and different additives that affect setting point times. Also time slicing reduces statistics, so repeated measurements are needed.

Dr Walewski remarked: What is the accuracy of DFT in this case?
Dr Chass replied: The DFT accuracy is acceptable but model sizes are poor.

Dr Drechsel-Grau communicated: How reliable is the model you use?
Dr Chass replied: The model is representative. It's based on similar trends for analogous materials.

5. Direct Measurement of Competing Quantum Effects on the Kinetic Energy of Heavy Water upon Melting (Giovanni Romanelli, Rome, Italy)

Prof Greaves opened the discussion of the lecture by Mr Romanelli: How do your calculations based on stretching modes compare with Lindemann's melting law where periodicity is lost when root-MSD approaches the size of interatomic distances?
Mr Romanelli replied: The interpretation of these experimental results is based on a harmonic potential and no dependence on the MSD is considered.

Prof Andreani remarked: Exploring MSD should be unimportant.

Dr Walewski commented: Why don't you present the bending mode contribution to the kinetic energy of a water molecule? Is it not available from the experiment?
Mr Romanelli commented in reply: The mean kinetic energy we measure is the sum of the contributions of all the rotations, translations and vibrations. The interpretation of the multivariate Gaussian fit is straightforward only for the component along the stretching modes, since in this direction the principal contribution comes from the most energetic modes. The motion along the other two directions is principally due to combination of rotations and bending, and then a model is needed in order to extract the information of the bending contribution from the directional kinetic energies.

Journal of Physics: Conference Series **571** (2014) 012005 doi:10.1088/1742-6596/571/1/012005

Epithermal neutron instrumentation at ISIS

G. Gorini[1], G. Festa[2,3] and C. Andreani[2,3]

[1]Università degli Studi di Milano-Bicocca, Milan, Italy
[2]Università degli Studi di Roma Tor Vergata, Rome, Italy
[3]CNR-IPCF, Messina, Italy
[4]CNR-IFP, Milan, Italy

Mail: giulia.festa@roma2.infn.it

Abstract. The advent of pulsed neutron sources makes available high epithermal neutron fluxes (in the energy range between 500 meV and 100 eV). New dedicated instrumentation, such as Resonance Detectors, was developed at ISIS spallation neutron source in the last years to apply the specific properties of this kind of neutron beam to the study of condensed matter. New detection strategies like *Filter Difference* method and *Foil Cycling Technique* were also developed in parallel to the detector improvement at the VESUVIO beamline. Recently, epithermal neutron beams were also used at the INES beamline to study elemental and isotopic composition of materials, with special application to cultural heritage studies. In this paper we review a series of epithermal neutron instrumentation developed at ISIS, their evolution over time and main results obtained.

Detectors for spectroscopy

The advent of pulsed neutron sources, such as the ISIS spallation source, has made available relatively large fluxes of neutrons in the *epithermal* energy range between 500 meV to 100 eV. Such energy range is usually not available in reactor-based neutron sources and even in pulsed facilities most of the beamlines are subdued to neutron moderators made to provide a beam of thermal (about 25 meV) neutrons. Only in recent times the possibilities given by epithermal neutrons have generated increased interest in the scientific community, due to the fact that neutrons in the lower part of the epithermal range provide information about atomic and molecular phenomena with characteristic energies of some eV, previously of difficult access through traditional probes. Consequently, new research paths for the investigation of certain aspects concerning condensed matter were opened, as for instance the short time single particle dynamics in quantum and molecular systems and high-energy excitations in materials. Since 1985 the ISIS spallation neutron source has been providing opportunities for research and development in experimental instrumentation and in new detection strategies to extend the kinematical region accessed by neutron scattering experiments to exchanged momenta $\hbar q \leq 10$ Å^{-1} and $\hbar q > 200$ Å^{-1} coupled with exchanged energies $\hbar\omega \geq 1$eV.
The VESUVIO spectrometer [1, 2] at ISIS was designed and built by British and Italian scientists for inelastic neutron scattering (INS) at such energy and momentum transfers ranges; it has been operational at the ISIS pulsed neutron source since September 2001. This instrument was an upgrade of the previous eVS spectrometer, that operate in the same energy range since 1985, pioneering inelastic neutron scattering at the eV energies in the last decades. The experimental technique

Journal of Physics: Conference Series **571** (2014) 012005 doi:10.1088/1742-6596/571/1/012005

implemented at VESUVIO is known as *Deep Inelastic Neutron Scattering* (DINS). The VESUVIO Spectrometer used the high intensity of neutrons in the eV energy range (about 10^6 n/s cm^2 eV) and the pulsed nature of the ISIS source to measure atomic momentum distributions in a variety of condensed matter systems [3]. When the momentum transfer in the scattering process is sufficiently large (at least 10 times higher than the mean atomic momentum), the scattering can be interpreted within the *Impulse Approximation* (IA) framework. In IA scattering is essentially a single-atom "billiard ball" scattering, entirely determined by conservation of momentum and *kinetic* energy in the collision (molecular effects in the sample are irrelevant). By measuring the energy and momentum changes of the neutron, the momentum component of the target atom along the direction of the scattering vector $\boldsymbol{q}$ can be determined. Energy transfers in the 1-30 eV region and wave-vector transfers between 30 A^{-1} and 200 A^{-1} are achieved using a *filter difference* (FD) technique. Being VESUVIO an inverted geometry spectrometer [4], the energy of the scattered neutron is fixed by a nuclear resonance absorption foil; the incident energy, and hence the energy and momentum transfers, are determined using standard time of flight (TOF) techniques. The nuclear resonance absorption foil is a basic element for eV neutron spectroscopy in an inverted geometry instrument, in that it selects the final energy of the scattered neutrons. The experimental method used at VESUVIO consists of cycling the analyzer foil in and out of the scattered neutron beam. In this way, two measurements are taken for every sample, one with the foil between sample and detector and one with the foil removed, as shown in figure 1_a.

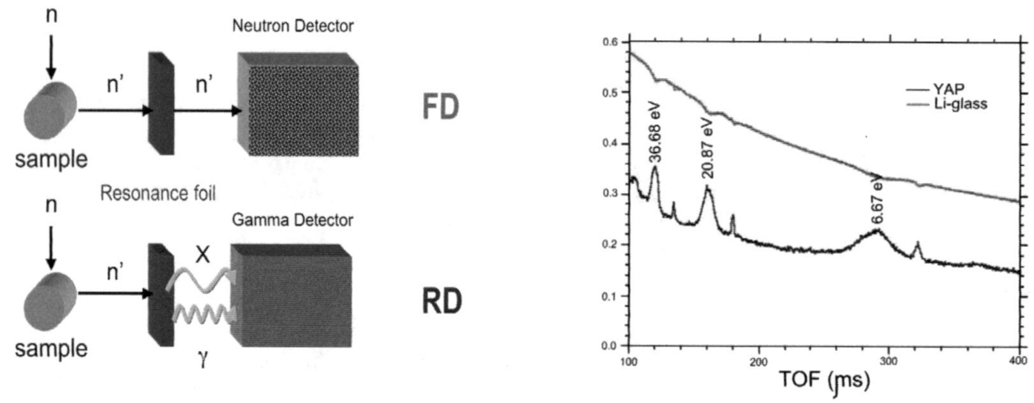

(a) (b)

Figure 1. (a) Principles of Filter Difference and Resonance Detector techniques. (b) Time of flight spectra for Filter Difference (Li-glass detector based) and Resonance Detector (YAP detector based) techniques.

The difference between these two data sets is indeed the experimental signal, i.e. a measurement of the intensity of neutrons scattered from the sample with final energy E_1. The upgraded VESUVIO spectrometer employs routinely ^{197}Au or ^{238}U foils as filter analyzers. The resonances of uranium are quite narrow compared to the resonance energy ($\Delta E_r/E_r$ ~0.5%), however the broadening due to the thermal motion of the lattice in which the absorbing nuclei are embedded, is an important contribution to the spectrometer resolution function; it was experimentally shown that the latter can be effectively reduced by cooling the filter foils.

Resonance Detectors (RD) were developed for VESUVIO and replaced the Lithium-glass detectors previously used on the eVS spectrometer [5, 6]. In the case of Resonance Detectors, the detection of the prompt γ-ray radiation emitted by the absorber foil after resonant neutron capture was used (figure 1_b). Neutron resonance radiative capture is a physical process which occurs in two steps: i) absorption of a neutron in a nucleus at a certain energy, mostly determined by the internal structure of the absorbing system and ii) prompt emission of a γ-rays cascade due to primary (from the absorbing nuclear energy level to the ground state) and multi-step radiative transitions [7]. The new RD set-up in VESUVIO was optimized and the performance of a new kind of photon detectors was tested including solid-state cadmium-zinc-telluride (CZT), a silicon detector based system, a Li-glass scintillator and yttrium aluminum perovskite (YAP) scintillators coupled with photomultipliers. The best configuration in terms of efficiency and signal to background ratio enhancement resulted to be YAP inorganic crystal coupled with a photomultiplier [8, 9, 10, 11]. YAP, indeed, is a fast, mechanically and chemically resistant scintillator material. Main characteristics of this material are a good light yield (about 18000 photons/MeV) and a short decay time of 27 ns for λ = 350 nm (wavelength of maximum emission). It is insensitive to neutrons and has a 100% efficiency at 100 KeV and 10% at 1 MeV. In figure 2 the TOF spectrum from an ice sample in an Aluminum cell in the time range 50-500 μs recorded through the YAP-based detector is shown as an example of the performances of the new system: details of the analysis of such spectrum are presented in ref. 12.

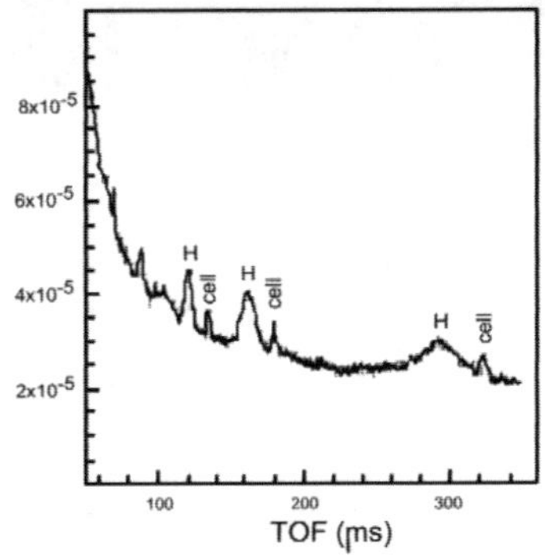

Figure 2. Time of flight spectra in the time range 50-500 μs recorded through the YAP based detector (sample: ice in an Aluminum cell). Four Hydrogen resonances are present and could be analyzed.

Research about Resonance Detectors improved the performance of the VESUVIO spectrometer by providing a high-resolution *Forward Scattering Detector* (FSD) bank and a *Very Low Angle Detector* (VLAD) bank (figure 3_a and 3_b). VLAD was designed to perform *High energy Inelastic Neutron Scattering* (HINS) measurements in the region of the kinematical space *(q, ħω)* characterized by $1.5\ \text{Å}^{-1} \leq q \leq 10\ \text{Å}^{-1}$ and $0.3\ \text{eV} \leq \hbar\omega \leq 20\ \text{eV}$, thus complementing the capability of the instrument beyond DINS measurement. The VLAD configuration equipment has proven to be effective up to final neutron energies above 100 eV [13, 14, 15]. On VLAD the YAP crystals are given a trapezoidal shape in order to cover a large area (about 1075 mm^2) in the same angular range (figure 3_c).

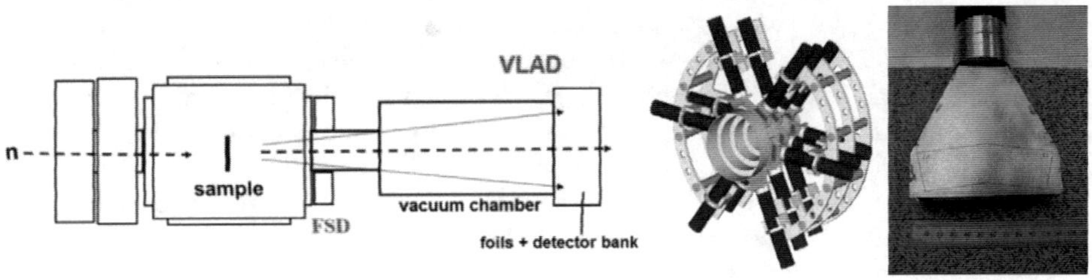

(a) (b) (c)

Figure 3. (a) VLAD and Forward Scattering Detector (FSD) set up on VESUVIO beamline; (b) scheme of VLAD detector bank; (c) VLAD detector: the YAP crystal is shaped in a trapezoidal and connection with a photomultiplier tube, through optical coupling by a light guide system.

Journal of Physics: Conference Series **571** (2014) 012005 doi:10.1088/1742-6596/571/1/012005

The particular crystal shape does not allow direct connection with a photomultiplier tube, so that the optical coupling of the active medium with the PMT is provided by a light guide system therefore ensuring adequate light collection. The FSD bank was also developed and installed on VESUVIO (figure_4).

Figure 4. Photograph of Forward Scattering Detector bank at VESUVIO beamline.

A preliminary set-up was made of two detector banks with 32 scintillators (eight rows of four scintillators each) placed on either side of the beam pipe. With such configuration the bank presented a noticeable cross-talk between detector elements and the isotropic emission of the gamma rays was a source of corruption of the angular information. To solve the issue an optimized set up was developed, increasing the distance between the scintillators and sixteen new modules with four scintillators each were mounted vertically (see figure_4). The FSD cover a range of angles from 33^0 to 67^0.

A further improvement to the beamline design was the application of a *Foil Cycling Technique* (FCT). The FCT uses two foils of the same neutron-absorbing materials; the first foil is placed in front of the gamma detector and is used as the energy *analyzer* for RD configuration; the second is the cycling foil that is placed between the sample and the analyzer. During measurements, the foils are cycled in and out the scattered neutron beam and two measurements are taken (*foil-in* and *foil-out*) combining the RD and the FD techniques. Final data are obtained as the difference between the two measurements.

1. Detectors for other applications: Neutron Resonance Transmission and Neutron Resonance Capture Analysis imaging techniques

Epithermal neutron beams can also be used to study materials in terms of elemental and isotopic composition identification. *Neutron Resonance Capture Analysis* (NRCA) uses the unique resonance absorption properties of epithermal neutrons. Many elements have neutron absorption resonances in the energy range below 500 eV. Their absorption cross-sections are well known and vary from one element to another and actually also between isotopes of the same element, thus a resonance is a sort of fingerprint of a chemical element and of its isotopes [16]. Neutron absorption is often followed by the prompt emission of a gamma-ray cascade, with typical total cascade energies up to about 8 MeV. Being a non-destructive, non-invasive technique, thus especially suited to samples that cannot be damaged, NRCA has been used for analysis of archaeological artifacts to determine elements' composition [16, 17-25].

In the framework of the *ANCIENT CHARM* (*Analysis by Neutron resonant Capture Imaging and other Emerging Neutron Techniques: new Cultural Heritage and Archaeological Research Methods*) project, funded by the European Community "New and Emerging Science and Technology" program, new imaging techniques based on improvements of the NRCA approach were developed at ISIS. A

Neutron Resonant Capture Imaging combined with Neutron Resonance Transmission (NRCI/NRT) set-up was installed on the INES beamline at ISIS (figure_5).

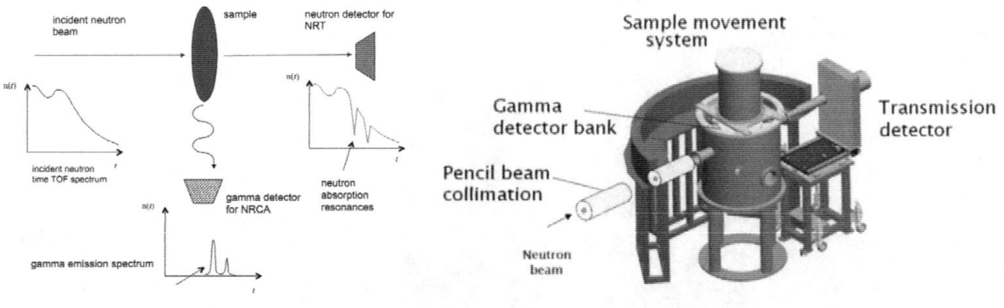

Figure 5. Neutron Resonant Capture Imaging combined with Neutron Resonance Transmission (NRCI/NRT) set-up installed at INES beamline, ISIS.

NRCI/NRT uses both gamma emission and neutron transmission measurements of resonant epithermal neutrons to determine the 3D elemental composition of the sample. The time of flight technique allows to recognize the resonance energies through the positions of the absorption peaks in the time spectrum. While the time position provides qualitative information (i.e. the nature of the element), the peak area provides quantitative information on the amount of the specific recognized element within the sample. Typical detection limits for NRCA are down to concentrations of 10^{-2}–10^{-4}%. In the case of NRT detection limits are worse around 10^{-1}–10^{-2}%.

The set-up for NRCI uses a 5 mm pencil beam of epithermal neutrons and an array of 28 YAP detectors mounted in a bank made of 3 support rings of cast lithium carbonate, surrounding the sample position. Each detector has a circular crystal front face with a 50 mm diameter and 25 mm depth (figure 6_a).

(a) (b)

Figure 6. (a) Neutron Resonant Capture Imaging detector array; (b) Neutron Resonance Transmission detector.

The whole gamma detector bank is installed inside the INES aluminum tank. The pencil beam collimation was obtained through two sections (400 mm long and 150 mm long), each made of 50% Li_2CO_3 and 50% epoxy resin and installed inside the INES shutter, approximately 3 m upstream from the sample position.

The NRT detector was installed in transmission. The transmission geometry allows, in principle, to use a position-sensitive neutron detector and a rotating/translating sample holder to achieve a full 3D scan of the object under measurement. The detector developed for Ancient Charm project (figure 6_b)

Journal of Physics: Conference Series **571** (2014) 012005 doi:10.1088/1742-6596/571/1/012005

consists of a 10x10 pixels array; each pixel is made of neutron-sensitive GS20 scintillator glass (6.6% total lithium of which 95% ^{6}Li) with an active area of 1.8 × 1.8 mm^2 and it is embedded in a 16 mm thick support made of boron nitride and separated by Al foils from neighboring pixels. The individual pixels are coupled via a 0.5 mm thick glass disperser to optical fibers which transport the light to photomultiplier tubes. Reference NRT measurements were performed on different samples to test the element-sensitivity of the imaging technique and some example images are shown in figure 7.

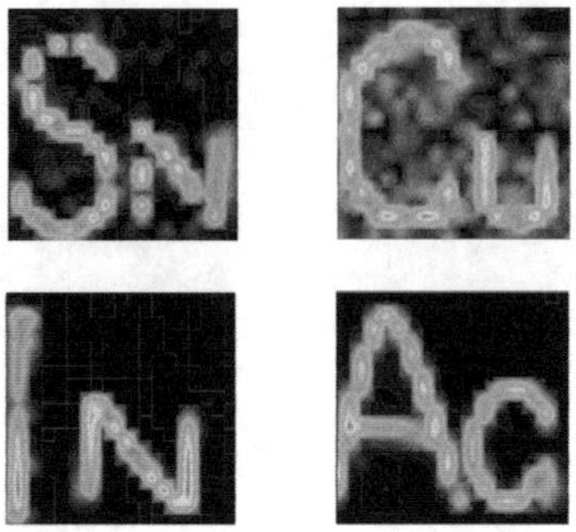

Figure 7. Spatial resolution in Neutron Resonance Transmission element sensitive imaging. Each square image has a side of 40 mm and the wires of different materials (tin-Sn, copper-Cu, indium-In, silver-Ag) are 1mm thick.

These are images of different 1 mm thick metal wires (tin-Sn, copper-Cu, indium-In, silver-Ag), superimposed and enveloped in a 1 mm thick Al foil. Wire shapes are clearly visible in the NRT images.

The NRCI/NRT set-up at INES was used on selected archaeological findings. In figure 8_a and 8_b a typical NRCA and NRT spectra are shown as an example of application to a realistic cultural heritage samples.

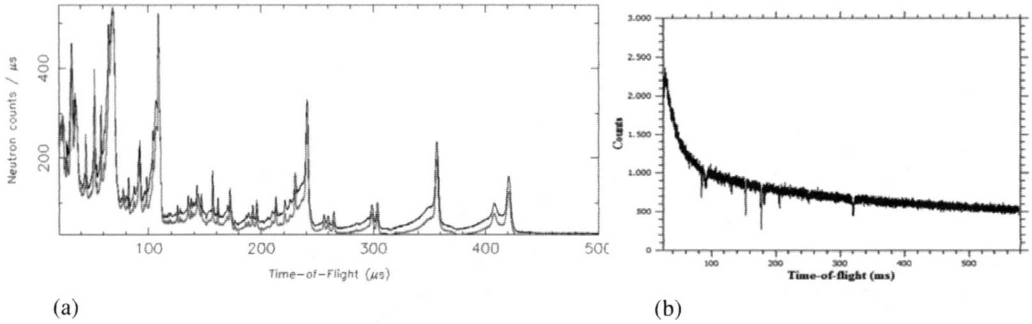

(a) (b)

Figure 8. (a) Spectrum from NRCA gamma detector of an archaeological interest sample; (b) Spectrum of NRT detector of an archaeological interest sample.

Shown spectra are related to the study of a gilded bronze head relief by the famous *maestro* Lorenzo Ghiberti. The sample comes from the East doors of the Florence Baptistery, also known as *'The Gates of Paradise'* for which information about the gilding state was obtained [26].

Conclusions and future prospective

An overview on how the availability of epithermal neutron fluxes has improved the development of dedicated neutron instrumentation at ISIS spallation neutron source has been provided. Epithermal neutron beams were used to study kinematical properties of matter and composition of materials in terms of both elemental and isotopic identification. In this framework, with the recent development of new instrumentation, new prospective is opened in different fields, for instance cultural heritage. Combined use of techniques like NRCI/NRT and the widely known Prompt Gamma-rays Activation Analysis [27] could increase - the possibilities of identifying "exotic" or unexpected elements and isotopes in Cultural Heritage samples. The detection of particular rare elements such as lanthanides or rare earths through mentioned imaging techniques could open a new generation of traceability strategies.

References

[1] C. Andreani, D. Colognesi, J. Mayers, G. F. Reiter, R. Senesi 2005 *Advances in Physics* **54** 377
[2] Mayers J and Reiter G 2012 *Meas. Sci. Technol.* **23** 045902
[3] Mayers J, Fielding AL, Senesi R 2002 *Nucl. Instrum. Methods A*, **481**(1-3) 454-463.
[4] Winsor C, *Pulsed Neutron Scattering*, London, UK; 1981
[5] Mayers J and Reiter G. 2012 *Meas. Sci. Technol*, **23** 045902 (18p)
[6] A. Pietropaolo, R. Senesi 2011 *Physics Reports* **508**, 45-90
[7] Postma H. and Schillebeeckx P., J. Radioanal. Nucl.Chem., 265 (2005) 297.
[8] Andreani C, D'Angelo A, Gorini G, Imberti S, Pietropaolo A, Rhodes NJ, Schooneveld EM, Senesi R, Tardocchi M. 2004 *Appl. Physics A* **78**, 903-913
[9] Tardocchi M, Pietropaolo A, Andreani C, Bracco A, D'Angelo A, Gorini G, Imberti S, Senesi R, Rhodes NJ, Schooneveld EM. 2004 *Nucl. Instrum. Methods A* **526** 477-492
[10] Tardocchi M, Gorini G, Pietropaolo A, Andreani C, Senesi R, Rhodes N, Schooneveld EM. 2004, *Rewiew of Scientific Instruments*. **75**, n.1 4880
[11] Tardocchi M, Arnaboldi C, Gorini G, Imberti S, Pessina G, Previtali E, Andreani C, Pietropaolo A, Senesi R. 2004 *Physics B* **350** e853-e856
[12] Tardocchi M, Pietropaolo A, Andreani C, Gorini G, Perelli-Cippo E, Rhodes NJ, Schooneveld EM, Senesi R. 2004 *Nucl. Instrum. Methods A* **535** 121-125
[13] Imberti S, Andreani C, Garbuio V, Gorini G, Pietropaolo A, Senesi R, Tardocchi M 2005 *Nucl. Instrum. Methods A* **552** 463-476
[14] Perelli-Cippo E, Gorini G, Tardocchi M, Cremonesi O, Andreani C, Pietropaolo A, Senesi R, Bowden ZA, Rhodes NJ, Schooneveld EM. 2008 *IEEE Trans. on Nucl. Sci.*, **52** n.4
[15] Pietropaolo A, Andreani C, D'Angelo A, Senesi R, Gorini G, Imberti S, Tardocchi M, Rhodes N, Schooneveld EM. 2002 *Appl. Phys A* ***74*** *[suppl]* S189-S1190
[16] Gorini G and Ancient Charm collaboration. 2007 *Nuovo Cimento*, **30**, N. 1.
[17] Postma H et al., 2001 *J. Radioanal. Nucl. Chem.*, **248**, 115.
[18] Mughabghab SF et al., Neutron Cross-sections,Vol. 1 (Academic Press, New York) 1981. Available online at http://atom.kaeri.re.kr/cgi-bin/endfplot.pl.
[19] Database for Prompt Gamma-ray Neutron Activation Analysis, Database developed as part of a Co-ordinated Research Project and sponsored by the International Atomic Energy Agency (IAEA), available online at http://www-nds.iaea.org/indew.html.
[20] Postma H et al., 2003 *Czech. J. Phys., Suppl. A*, **63** A233.
[21] Blaauw M, Postma H and Mutti P 2005 *Appl. Radiat. Isotopes*, **62** 429.
[22] Blaauw M, Postma H and Mutti P, *Nucl. Instrum. Methods A*, **505** (2003) 508.
[23] Postma H, Schillebeeckx P and Halbertsma RB, *Archaeometry*, **46** (2004) 635.
[24] Postma H, Perego RC, Schillebeeckx P, Siegler P and Borella A 2007 *J. Radioanal. Nucl. Chem.*, 271 (2007) 95.
[25] Postma H et al., Lunula. Archaeologia protohistorica, XIII (2005) 41.
[26] Perelli Cippo E, Borella A, Gorini G, Kockelmann W, Mox M, Postma H, Rhodes NJ, Schillebeeckx P, Schoonenveld EM, Tardocchi M, Dusz K, Hajnal S, Biro K, Porcinai S, Andreani C and Festa G 2011 *Jour. of Anal. Atom. Spectr.*, **26**, 992.
[27] Molnar GL, Handbook of Prompt Gamma Activation Analysis with Neutron Beams. *Kluwer Academic Publishers*, 2004.

Journal of Physics: Conference Series **571** (2014) 012006 doi:10.1088/1742-6596/571/1/012006

The VESUVIO Spectrometer Now and When?

A.G. Seel[1,2*], **M. Krzystyniak**[1,3], **and F. Fernandez-Alonso**[1,4]

[1] ISIS Facility, Rutherford Appleton Laboratory, Chilton, Didcot, Oxfordshire OX11 0QX, United Kingdom

[2] Department of Chemistry, University of Oxford, South Parks Road, Oxford OX1 3QR, United Kingdom

[3] School of Science and Technology, Nottingham Trent University, Clifton Campus, Nottingham, NG11 8NS, United Kingdom

[4] Department of Physics and Astronomy, University College London, Gower Street, London WC1E 6BT, United Kingdom

E-mail: andrew.seel@stfc.ac.uk

Abstract. The current layout and mechanics of the VESUVIO spectrometer are presented in light of spectroscopic measurements using electron-volt neutrons. A brief background to the theoretical framework of deep inelastic neutron scattering is presented, with focus on data collection and instrumental design. The current capabilities and research themes for VESUVIO are discussed, and possible future instrumental developments highlighted which will enhance the instrument's ability to meet scientific inquiry and expectation.

1. Introduction

The VESUVIO spectrometer at the ISIS Pulsed Neutron and Muon Source (Rutherford Appleton Laboratory, UK) is an instrument that probes a unique kinematic region in neutron scattering [1]. Despite there now being a number of spallation neutron sources across the globe, VESUVIO has remained the only spectrometer routinely serving a user community for measurements of scattering processes utilising high-energy neutrons in the epithermal region (5-150 eV), termed Deep Inelastic Neutron Scattering (DINS). In recent years the Bariloche Electron LINAC (Argentina) has witnessed the implementation of a spectrometer for DINS measurements. Whilst VESUVIO continues as both forerunner and standard for the measurement of atomic momentum distributions in condensed matter research, and remains in active development, it is hoped that further development in this field will see the advent of additional instruments.

In this article, the current layout and brief theoretical background for VESUVIO are presented, alongside the mechanics of a VESUVIO measurement as would be suitable for a prospective user of

* To whom any correspondence should be addressed.

Journal of Physics: Conference Series **571** (2014) 012006 doi:10.1088/1742-6596/571/1/012006

the instrument. In this manner, we hope to give a simple overview of the current state and operation of the instrument, highlighting both themes and capabilities in the field of electron-volt neutron scattering. Following this, there will be presented a future outlook for DINS and a discussion of how VESUVIO could be developed further to both meet the evolving-requirements of its user community, and to impact new fields of research.

2. Neutron Compton Scattering and a VESUVIO measurement

1.1. The Current Layout of VESUVIO

VESUVIO, as with the majority of spectrometers on a pulsed neutron source, is a time-of-flight (ToF) spectrometer, with ToF value, *t*, for each detected neutron given by [1,2]

$$t = \frac{L_0}{v_0} + \frac{L_1}{v_1} \tag{2.1}$$

Where L_0 and L_1 define the distance of sample from source and detector respectively, v_0, is the incident neutron velocity and v_1 the velocity of detected neutron. The neutron velocity can be related to its energy, E, through the relation $v = \sqrt{2E/m}$ where m is the neutron mass. As such, and through knowledge of t, L_0, L_1 and either E_0 or E_1, the energy transfer, $\hbar\omega$, and momentum transfer, q, can be determined [2]

$$\hbar\omega = E_0 - E_1 \tag{2.2}$$

$$q = \sqrt{2m}\left(E_0 + E_1 - 2\sqrt{E_0 E_1}\cos\theta\right)^{1/2} \tag{2.3}$$

VESUVIO operates in inverse geometry, and as such E_1 is fixed at 4897meV.

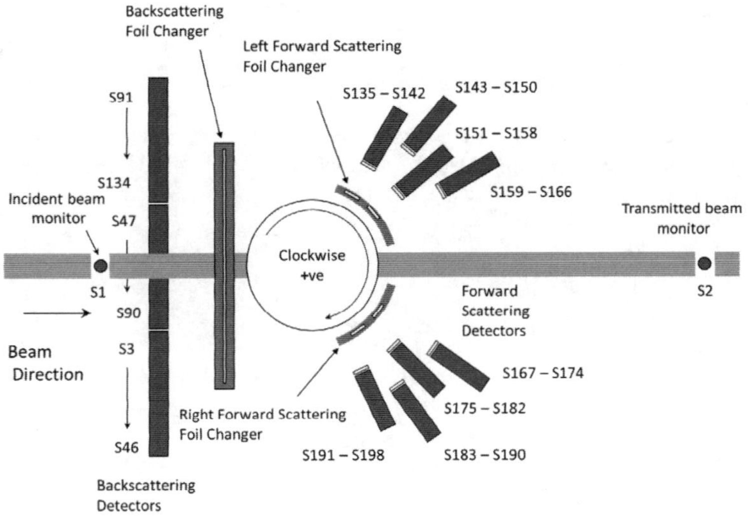

Figure 1. Schematic layout of the VESUVIO spectrometer

The layout of VESUVIO is given in figure 1. The instrument comprises 196 detectors in addition to an incident (S1) and transmitted (S2) beam monitor. It can be seen that two distinct geometries for detector banks exist, those in forward scattering (S135-S198) and backscattering (S3-S134). In backscattering the detectors are laid out in three banks perpendicular to the beam direction. These Li-doped glass scintillator detectors cover an angular range of 130°-163°, with each detector of approximate area 20 cm^2.

E_I is defined and fixed on VESUVIO through the incorporation of gold analyser foils. In backscattering this is done by taking measurements with and without foils between detector and sample [3]. Gold exhibits a strong nuclear resonance, absorbing neutrons at an energy value calibrated as 4897meV. Thus a difference measurement will yield ToF spectra corresponding to scattering processes resulting in this fixed E_I. By taking measurements of different gold foil thicknesses, the energy resolution of VESUVIO is improved. This 'Double Difference' (DD) foil cycling technique is detailed elsewhere [1,3-4]. It should be noted that running VESUVIO in 'foil-out' mode, *i.e.* without analyser foils present, allows for the instrument to be utilised as a diffractometer. Foil-cycling the 132 ^{6}Li-doped scintillator detectors in backscattering geometry on VESUVIO thus allows for simultaneous diffraction measurements in the thermal neutron region during a standard DINS measurement.

The detector banks at forward scattering on VESUVIO were upgraded in 2008, now consisting of γ-detecting Yttrium Aluminium Perovskite (YAP) detectors [5]. The gold resonance at 4897 meV is still used to define E_I, but in this case it is the ejected γ-ray that is detected. The DD technique is instigated through the use of the secondary gold foils detailed in figure 1. The cycling of these secondary analyser foils produces a sample-independent γ-background of ~5% in forward scattering [6]. This unwanted contribution to the signal is subtracted from spectra as part of the data analysis process.

1.2. Atomic Momentum Distributions and the Impulse Approximation
The aim of any measurement on VESUVIO is to probe the atomic dynamics of a system [7], through determination of atomic momentum distributions or kinetic energies. The reduction and analysis of raw data is viewed by the user in time of flight, *C(t),* or in a given, mass-specific response function in *y*-space $J_M(y)$ (rather than the scattering law S($\boldsymbol{q}$,ω) more common for other neutron spectrometers), and so in this section we shall briefly outline the mechanics of VESUVIO data handling within the framework of DINS.

At the high energy transfers available on VESUVIO, the scattering event is described within the Impulse Approximation (IA), whereby conservation of momentum and kinetic energy occurs between the incident neutron and single target nucleus. As such, the scattering law may be written as [7-9]

$$S(q,\omega) = \int n_M(\boldsymbol{p})\delta\left(\omega + \frac{p^2}{2M} - \frac{(\boldsymbol{p}+\boldsymbol{q})^2}{2M}\right)d\boldsymbol{p} = \frac{M}{q}J_M(y_M,\widehat{\boldsymbol{q}}) \tag{2.4}$$

where

$$J_M(y_M,\widehat{\boldsymbol{q}}) = \int n_M(\boldsymbol{p})\,\delta(y_M - \boldsymbol{p}\cdot\widehat{\boldsymbol{q}})d\boldsymbol{p} \tag{2.5}$$

and

$$y_M = \frac{M}{q}\left(\omega - \frac{q^2}{2M}\right) \tag{2.6}$$

Journal of Physics: Conference Series **571** (2014) 012006 doi:10.1088/1742-6596/571/1/012006

where $n_M(\boldsymbol{p})$ is the radial nuclear momentum distribution for species of atomic mass M, and $J(y,\boldsymbol{q})$ thus being the distribution of momentum states along the direction $\widehat{\boldsymbol{q}}$ $(= \boldsymbol{q}/q)$, the Compton profile. It is immediately apparent that within the IA there is a mass dependence, with the transform to y-space centring the contribution from each atomic species at y=0, dependent on the recoil energy for a given mass M, $q^2/2M$. This is expressed in the count rate for VESUVIO spectra, given by

$$C(t) = \frac{E_0 I(E_0)}{q} \sum_M A_M M J_M(y_M) \otimes R_M \tag{2.7}$$

where $I(E_0)$ is the incident flux of neutrons with energy E_0. The mass-dependent amplitude factor, A, is proportional to the number density (N_M) and scattering cross-section (σ^T) of the atomic species, and R is the mass-dependent instrument resolution function.

A DINS measurement on VESUVIO (for a given detector geometry) thus consists of a set of peaks in ToF, centred at values relating to each atomic species' recoil energy, itself dependent on the atomic mass. In practise a majority of experiments on VESUVIO are on isotropic samples, whereby the $\widehat{\boldsymbol{q}}$ dependence of $J(y)$ is lost. The most common model for $J(y)$ is a Gaussian form in y, exact for an isotropic, harmonic potential. In this case the atomic kinetic energy $\langle E_K \rangle$ relates to the standard deviation of the Gaussian, σ, through

$$J_M(y_M) = \frac{1}{\sqrt{2\pi\sigma_M^2}} \exp(-y_M^2/2\sigma_M^2) \tag{2.8}$$

$$\langle E_K \rangle = \frac{3\hbar\sigma_M^2}{2M} \tag{2.9}$$

This simple expression represents the elegance of DINS. For a given system, each mass is expressed as an individual peak in ToF, the widths of which (in actuality the standard deviation of Gaussian peak in the y-space for each mass) are a direct measurement of their mean kinetic energy and thus confining potential. Further models for $J(y)$ that move beyond the isotropic harmonic-oscillator model include expansion of equation (2.8) to include final-state effects [7,9], higher-order Hermite polynomials [7,10] (the so-called Gram-Charlier expansion), and the use of a multivariate Gaussian distribution [11].

3. Current Scientific Directions and Future Developments for VESUVIO

1.3. The Problem of Mass

It has been stated that in a DINS measurement each mass is exhibited as a peak centred at its recoil energy (in ToF this is dependent on instrument geometry), with intensities related directly to number density and scattering cross-section. As mass is increased, the separation between neighbouring elements becomes progressively shorter, as can be seen from equations (2.4-2.7). It is therefore a simple matter to separate and compare the lightest masses, such as hydrogen, deuterium (D) or lithium (Li), but for masses such as carbon (C) and nitrogen (N) there is swift convergence. Figure 2 demonstrates this convergence, and highlights the difference in convergence rates in forward and backscattering geometries.

As such, for measurements whereby masses >5 amu are of interest, and particularly for systems containing multiple masses which are desired to examine simultaneously, it is the backscattering banks that are of interest. By verity of being (nearly) isobaric with the neutron, hydrogen will not backscatter in the epithermal region, and thus a peak arising from neutron scattering (strictly we confine this to a single scattering event) from a proton will only appear in forward scattering spectra. We can thus divide discussion of the scientific capabilities (or directions) of VESUVIO into two thematic areas, those involving the proton and deuteron, and those of simultaneous measurement of heavier masses.

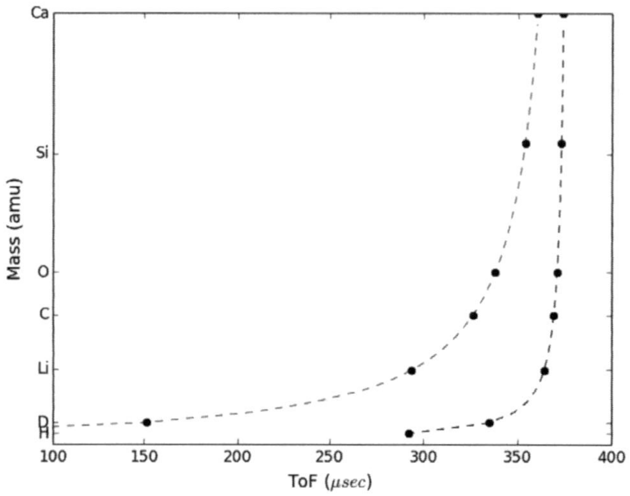

Figure 2. Recoil peak positions for atomic masses 1.0079-40.0 amu. Detector geometries simulated in forward (blue) and backscattering (green) for θ=40.0° and 140.0° respectively, with L_1=0.52m. These are close to the values for detectors S69 and S156.

1.4. Nuclear Quantum Effects

The last decade has seen a dramatic change in the science conducted on VESUVIO. An examination of the number of days requested on the instrument over recent cycles (figure 3) shows that the proton is the focus of a majority of research on VESUVIO. Increasingly more complex systems are being measured, whereby not only proton dynamics are important, but also nuclear quantum effects associated with the exchange of the proton (H) by the deuteron (D). Importantly, there is increasing interest to measure systems involving both H and D.

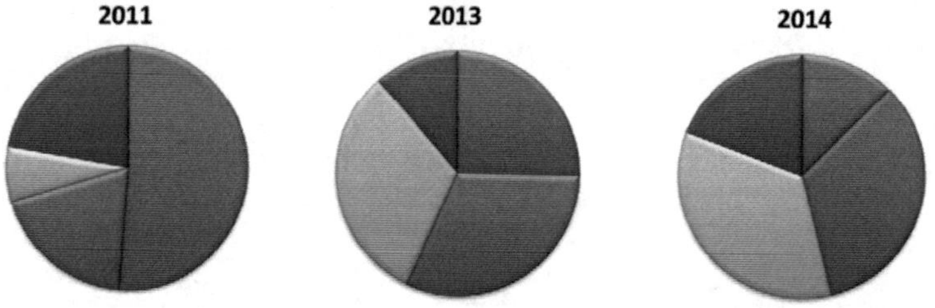

Figure 3. Recent requested time on VESUVIO divided into research themes. *Red*: H_2O, *Blue*: investigations into proton dynamics excluding H_2O; *Purple*: deuterium dynamics; *Green*: M>5 amu.

The large total scattering cross section of the proton combined with its light mass results in the proton peak being of particular high intensity and isolated from others in DINS spectra. As such VESUVIO has proven particularly adept in examining proton kinetic energies and momentum distributions in a variety of condensed matter systems [7], including diatomic H_2 [12,13], energy materials [14-16] and H_2O both in bulk states and confined within host materials [17-19].

Recent investigations into nuclear quantum effects, specifically those involving H/D comparison have all involved the measurement of H/D species separately [20,21]. Whilst not an outlandish approach by any means, this is primarily limited to systems with single proton environments. If VESUVIO and by extension DINS are to contribute to systems of partial deuteration, a greater capability in the measurement of both H and D simultaneously must be achieved. In principle, this is something relatively simple on VESUVIO, with the mass-specific nature of the scattering response yielding signals from H and D separately in ToF. However, the sparseness of VESUVIO data (by verity of the low detected flux on the instrument) combined with the often far larger H>D signal can make reliable measurement of D kinetic energy difficult. This is exemplified in the simulated spectrum for monodeuterated/monohydrogenated phenol, shown in figure 4.

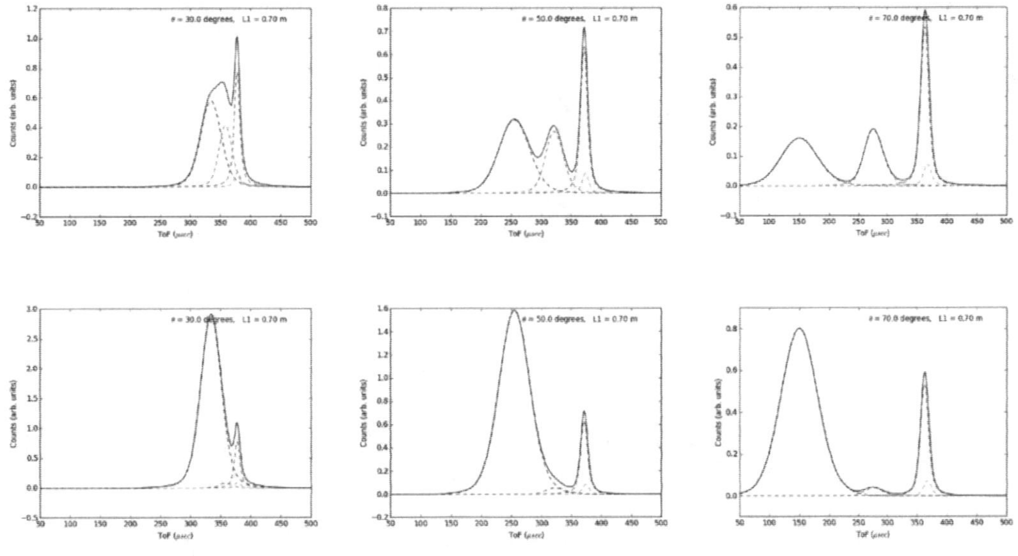

Figure 4. Simulated DINS spectra for isotopically labelled phenol for L_1=0.7 m and θ=30° (*left*), 50° (*centre*) and 70° (*right*). *Top row*: C_6D_5OH. *Bottom row*: C_6H_5OD.

VESUVIO is currently flux-limited, particularly for the traditional, detector-by-detector approach to data reduction and analysis. An average measurement of the proton momentum distribution of a single sample at one temperature takes of the order of 18 hours. After this time the signal-to-noise levels for single-detector spectra is suitable for analysis of peak width and shape. Examining the simulated (and noiseless) spectra for C_6H_5OD it is clear that simultaneous fitting of the D peak is only possible for higher angles, $\theta \gtrsim 50°$.

In principle, measurement of D-profiles is possible in backscattering geometry, where the proton recoil signal is excluded from the kinematics of proton-neutron collisions using epithermal neutrons. It thus seems natural that for samples dominated by proton scattering the backscattering banks could be utilized in the isolation of D from H. Unfortunately, the current geometry of the VESUVIO banks yields a D-signal at ToF values <150μs, a region suffering from low signal-to-noise for the ^{6}Li-doped scintillator detectors in backscattering. Were the D-peak brought up to higher ToF values by shifting to lower angles in backscattering, routine measurements of D on these banks would be feasible. As such, we can conclude that VESUVIO is not currently exploiting two important angular ranges for simultaneous measurement of H/D – higher angles in forward scattering and lower angles in backscattering. These regions are shown in figure 5. An alternative to this would be to extend L_0 by physically moving the instrument further from the neutron source. This is currently not feasible at the operational levels of ISIS due to the concomitant loss of flux this would entail.

1.5. Heavier Masses and MANSE

Perhaps the most striking change in the scientific direction of VESUVIO shown in figure 3 is that 'heavy-mass' measurement, by which we refer to masses >5 amu, those of boron and above. Focus has been brought to the theoretical framework of DINS as a measure of atomic dynamics as separable by mass, now termed MAss-selective Neutron SpEctroscopy, or MANSE in the literature [22,23]. The advent of MANSE has been recent enough that the capabilities of VESUVIO in this field are yet to be explored to its full extent, with fresh approaches not only to sample chemistry/stoichiometry, but also data analysis and sample environment. Here we shall briefly comment on these, focusing on the manner in which VESUVIO could be developed in the future to accommodate MANSE in its scientific repertoire.

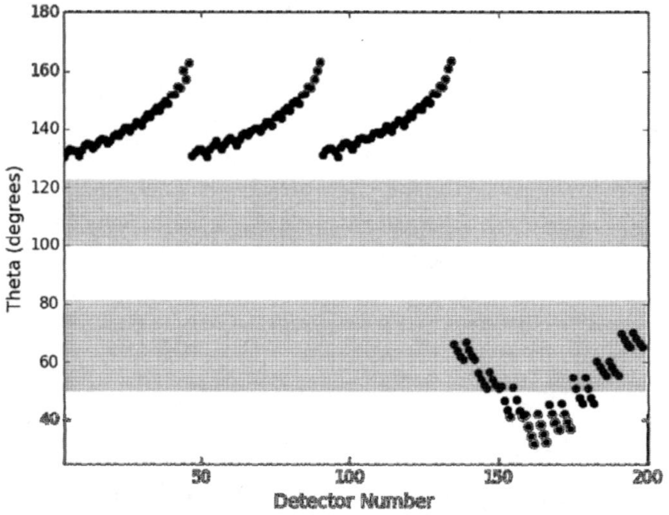

Figure 5. Angular coverage of detectors on VESUVIO in backscattering (3-134) and forward scattering (135-198). Shaded regions represent the angular coverage that could be exploited to further optimise VESUVIO for deuterium studies.

Examining again figure 2 we can see that for elements in the first two periods of the periodic table, H-Ne, their recoil positions are not yet converging in backscattering to an extent that is comparable to forward scattering. Of course this simplistic approach does not yet incorporate peak intensities or widths in ToF, but it is a suitable first estimation for the capability of VESUVIO to accurately and simultaneously measure the atomic kinetic energies of these elements in a given system. Within this range exist elements particularly important for modern materials and chemical research, for example lithium (battery and energy storage materials), boron (superhard materials), carbon and nitrogen (organic molecules and layered materials), oxygen (oxide devices) and fluorine (superacids and ion conductors). These serve only as examples, but highlight the potential breadth of application for DINS and VESUVIO.

As with forward scattering, the problem of flux and signal-to-noise is encountered in MANSE. Noiseless spectra for elements from Li to F are simulated in figure 6, with arbitrary values for profile width, σ_M. Now it can be seen that despite peak centroid separation in ToF, appreciable overlap in signal will result in ambiguity in fit parameters if data is sparse, as is the case for the low detected flux on VESUVIO. In addition to this, the sample environment itself contributes strongly to measurements of heavier masses. It is common to run samples within Al sample containers for neutron spectroscopic measurements, due to its low total scattering cross-section, 1.5 barn. This is equivalent to many elements of interest in DINS, and is indeed higher than some (σ^T_{Li}=1.37, σ^T_O=4.232 barn), and thus for smaller amounts of sample this contribution can dominate. This is demonstrated in figure 6, with measured spectra for a 1-mm-thick Li_2O sample run in a standard VESUVIO Al sample container compared to that simulated for a Sn container. Clearly the contribution from sample container can be reduced not only in absolute scattering power (by moving to thinner sample containers) but also by utilizing materials heavier than Al. For this reason, candidates include Sn-foils for standard and low-temperature measurements, and Nb containers for higher temperatures.

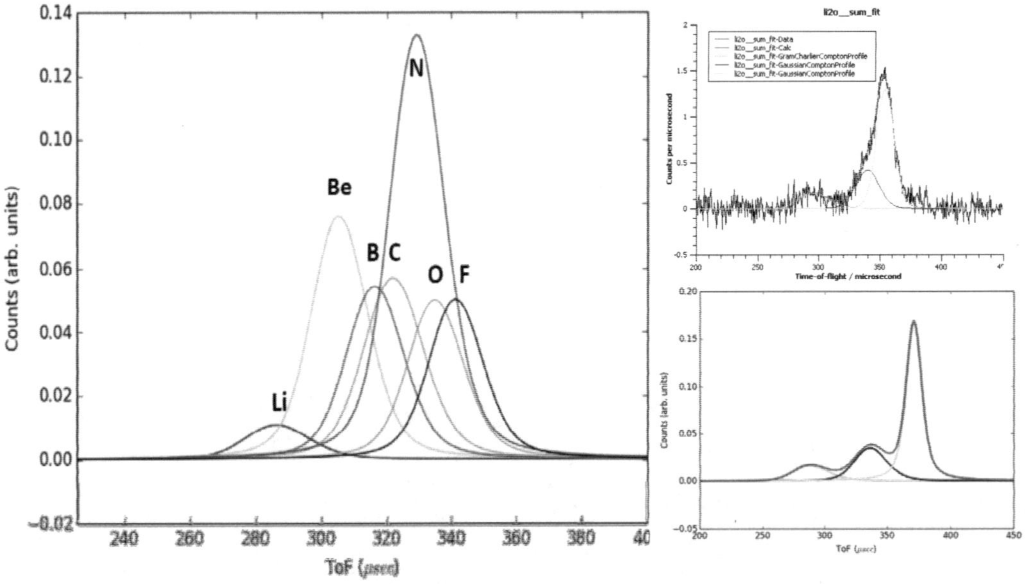

Figure 6. *Left:* Simulated DINS spectra for elements Li to F, L_1=0.55 m θ=140°. *Right Top*: Measured Li_2O spectra in standard Al sample container. *Right Bottom*: Simulated Li_2O in Sn container of equal scattering power.

Traditional analysis of DINS spectra on VESUVIO have been performed using a detector-by-detector methodology [1,24]. In this manner, the $J(y)$ for each mass is constructed in their individual y-spaces, with the composite function fitted to the ToF spectra for each detector, utilising the individual geometries for each. In this manner it follows that from each detector an independent σ_M is fitted. This historical perspective arises from the first measurements of σ_M for light elements, testing the impulse approximation and invariance of σ with detector angle through conversion to y-space.

But is this necessary? For the sparse data on VESUVIO, any means that can increase the statistics of the ToF signal for the purpose of spectral fitting, without degrading the signal, would be beneficial. An immediate solution would be to sum over detectors within a narrow $d\theta dL_1$ range. In this manner, if the difference in the ToF expression of *J(y)* (which we can simplify as a difference in centroid positions and peak widths ($dC, d\Gamma$) is smaller than the error in fitted σ_M (in ToF) for individual detectors, a fit to the summation of these detectors will not yield spurious σ_M values. There is also the possibility of a great saving in measurement time, whereas current runs for a single temperature take upwards of 18 hours. A visual representation for this is given in figure 7, where it can clearly be seen that spectra condense at lower and higher angles (128°-132°, 158°-162°). In practise it would therefore be feasible to define artificial detector banks within the backscattering detector modules, whereby ToF spectra for detectors of similar θ and L_1 values are summed and fitted to a single C(t) of averaged detector geometry. As discussed by Krzystyniak *et al.* in the same issue, a way to increase the information content of sparse data on VESUVIO involves an examination of so-called Cumulative Angle-Averaged Data (hereafter CAAD) in either forward or backscattering geometry. In this approach, scattering data measured at a given detector is first integrated in the ToF domain using numerical integration, then normalized by its TOF integral, and finally summed over a given range of detectors with the CAAD fitting function following the same fate as the CAAD signal. This data-analysis procedure leads to a more transparent assessment of the performance of a given nuclear-momentum-distribution model, as well provides a convenient starting point for the development of the 'detector-focusing' schemes described above.

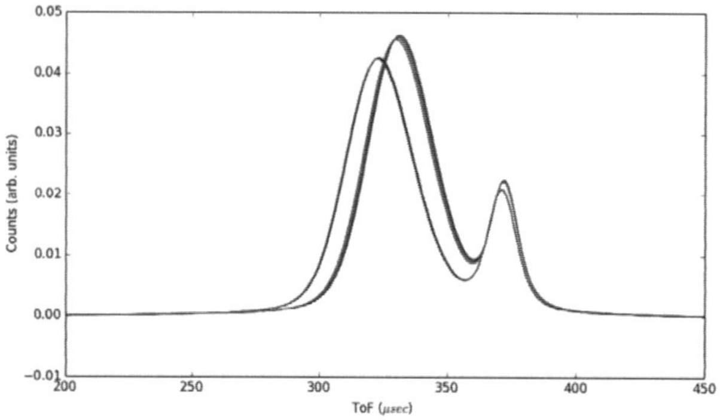

Figure 7. Simulated DINS spectra for C in Sn-foil. The overlaid spectra are for θ=128.0,129.0,130.0, 131.0 and 158.0,159.0,160.0,161.0,162.0.

4. Conclusions: VESUVIO When?

The VESUVIO spectrometer remains the only instrument providing a user programme for DINS research. Since its last major upgrade, a great advancement and diversity in the science performed by the instrument has been witnessed, as demonstrated by the research themes underpinning beam-time requests. We have proposed a number of instrument upgrades and extensions to VESUVIO that would enhance the current experimental capabilities of the instrument.

The most important limitation of VESUVIO is not one of design, but of flux. Were an upgrade to the ISIS target and moderators to be undertaken that would result in a higher incident flux (and thus count-rate) on VESUVIO, the effects on scientific capability cannot be overstated. Experimental durations would be brought down from the current 18 hours for routine measurements to those of inelastic or quasielastic neutron scattering. Increased flux would also enable an increase in L_0, which could vastly improve mass discrimination for H/D and MANSE experiments. This flux would currently come at a cost for the existing backscattering detectors, due to saturation effects. The ^{6}Li-doped glass scintillator detectors in backscattering are limited in terms of their electronics, in that they will saturate at approximately 45% scattering under the current flux on the instrument. Pronounced Bragg scattering from either sample or sample container can thus affect backscattering performance. An increase in flux on VESUVIO would either need to be attenuated in backscattering to avoid detector saturation or be compensated by an increase of L_1 (thus negating any possible gains on the instrument). Alternatively, replacing these detectors with the γ-detection technology already existing in forward scattering, where saturation is not a problem, would overcome this problem all together. The installation of a specific diffraction bank (scintillator detectors in backscattering at larger L_1) would still enable crystallographic identification in conjunction in this case.

There is a great need to extend the angular-coverage and number of detector banks on VESUVIO. By concentrating on higher forward-scattering and lower backscattering regions (a stark gap in the current instrument design) isolation and measurement of the kinetic energy and momentum distribution of D could be undertaken *simultaneously* to H. This extends DINS to a range of partially-deuterated samples that are currently difficult to measure on VESUVIO, but that are potentially important for the isolation and examination of nuclear quantum effects.

The manner in which MANSE samples are both measured and analysed is being revisited. Sample environment, particularly sample containers) plays a far greater role in these measurements than for those of the proton. We have demonstrated that by moving to foils of heavy-elements (Sn, Nb) rather than the Al containers commonly used, environment-intensity is shifted away from masses of interest in MANSE studies. The detector-by-detector methodology for DINS is also not optimised for MANSE, due to sparse, overlapping-intensity from masses >5 amu. One alternative proposed here is the approximation of average detector fitting. We propose that the error inherent in fitting to an averaged $C(t)$ is negligible for heavier masses, and far out-weighed by the improved statistics of summation.

Finally, one point should be mentioned that has been highlighted previously for future upgrades of VESUVIO [25,26], and that is the redesign of vertical detector- geometry in forward scattering. The resolution of VESUVIO for the measurement of both H and D is unparalleled, approximately 25% and 20% of the intrinsic width in y-space respectively. Of this, angular resolution dominates for H and is significant for D. By moving to a conical detector array around a Debye-Scherrer cone relative to the transmitted-beam axis, not only would the angular resolution for these detectors be improved by a factor of approximately 2, but the γ-background from the secondary analyser foils would be averaged to zero. This would be particularly important for increases in flux whereby mitigation of this sample independent γ-background would become relevant. If these detectors were also moved to a longer L_1 (again increasing the angular resolution for the detectors), the capabilities of VESUVIO for H/D momentum profiles would be drastically improved. This is, again, dependent on an increased flux.

References

[1] Mayers J and Reiter G 2012 *Meas. Sci. Technol.* **23** 045902
[2] Mayers J and Abdul-Redah T 2004 *J.Phys. Condes. Matter* **16** 4811-32
[3] Schoonveld EM *et al* 2006 *Rev. Sci. Instrum.* **77** 095103
[4] Seeger PA Taylor AD and Brugger RM 1985 *Nucl. Inst. Methods A* **240** 98
[5] Tardocchi, M *et al* 2004 *Rev. Sci. Inst.* **75** 4880

[6] Mayers J 2011 *Meas. Sci. Technol.* **22** 015903
[7] Andreani C, Colognesi D, Mayers J, Reiter G and Senesi R 2005 *Adv. Phys.* **54** 377-469
[8] Lovesey SW 1984 *Theory of Neutron Scattering from Condensed Matter* vol 1 (Oxford University Press)
[9] Sears VF 1984 *Phys. Rev. B* **30** 44
[10] Reiter GF and Silver R 1985 *Phys. Rev. Lett.* **54**, 1047
[11] Flammini D et al 2012 *J. Chem. Phys.* **136** 024504
[12] Andreani C, Colognesi D and Pace E 1999 *Phys Rev. B* **60** 10008-10021
[13] Mayers J 1993 *Phys. Rev. Lett.* **71** 1553-1556
[14] Krzystyniak M *et al* 2011 *Faraday Disscuss.* **151** 171-197
[15] Pietropaolo A *et al* 2012 *J. Chem. Phys.* **137** 204309
[16] Reiter GF, Mayers J and Platzman P 2002 *Phys. Rev. Lett.* **89** 135505
[17] Reiter GF, Mayers J and Noreland J 2002 *Phys. Rev. B* **65** 104305-104313
[18] Pantalei C *et al* 2008 *Phys. Rev. Lett.* **100** 177801
[19] Reiter GF *et al* 2012 *Phys. Rev.* B **85** 045403
[20] Giuliani A, Ricci MA, Mayers J and Bruni F 2012 *Phys. Rev. B* **86** 104308
[21] Romanelli G et al 2013 *J. Phys. Chem. Lett.* **4** 3251-3256
[22] Seel AG et al 2012 *J. Phys. Condens. Matter* **24** 365401
[23] Krzystyniak M *et al* 2013 *Phys. Rev. B* **88** 184304
[24] Mayers J 2012 *User guide to VESUVIO data analysis programs for powders and liquids* (Rutherford Appleton Laboratory)
[25] Mayers J and Adams MA 2009 *Calibration of the electron volt spectrometer VESUVIO at ISIS* (Rutherford Appleton Laboratory Technical Reports) **RAL-TR-2009-022**
[26] Mayers J and Adams MA, 2011, *Nucl. Instrum. Meth. A*, **625**, 47-56

Measurement of proton momentum distributions using a direct geometry instrument

R. Senesi[1,2], A. I. Kolesnikov[3], C. Andreani[1,2]

[1] Università degli Studi di Roma Tor Vegata, Dipartimento di Fisica and Centro NAST, via della Ricerca Scientifica 1, Rome 00133, Italy

[2] CNR-IPCF Sezione di Messina, Messina, Italy

[3]Chemical and Engineering Materials Division, Oak Ridge National Laboratory, Oak Ridge, Tennessee 37831, United States of America

E-mail: roberto.senesi@uniroma2.it

Abstract. We report the results of inelastic neutron scattering measurements on bulk water and ice using the direct geometry SEQUOIA chopper spectrometer at the Spallation Neutron Source (U. S. A.), with incident energy E_i= 6 eV. In this set up the measurements allow to access the Deep Inelastic Neutron Scattering regime. The scattering is centred at the proton recoil energy given by the impulse approximation, and the shape of the recoil peak conveys information on the proton momentum distribution in the system. The comparison with the performance of inverse geometry instruments, such as VESUVIO at the ISIS source (U. K.), shows that complementary information can be accessed by the use of direct and inverse geometry instruments. Analysis of the neutron Compton profiles shows that the proton kinetic energy in ice at 271 K is larger than in room temperature liquid water, in agreement with previous measurements on VESUVIO.

1. Introduction

Recent developments of the Deep Inelastic Neutron Scattering (DINS) technique for the measurement of atomic momentum distributions in materials provide an insight into the local environment of light (proton, deuteron, helium, lithium) and intermediate-weight (oxygen, fluorine, sodium, etc.) elements [1–10] in a variety of settings. The broader scope of these experiments is to directly measure the binding, anisotropy, and anharmonicity of the local potential experienced by the struck atoms. Isotropic quantum harmonic oscillator motion, although being the foundation for understanding diverse phenomena such as the vibrational modes of molecules, the motion of atoms in a lattice and the theory of heat capacity, is uncommon in materials such as metal hydrides and hydrogen-bonded systems, due to the presence of anisotropy and anharmonic effects. Indeed, the observation of atomic motion with nearly "perfect" quantum harmonic oscillations is regarded as an unexpected discovery, such as the recent Inelastic Neutron Scattering (INS) and DINS investigations of the quantum oscillations of nitrogen atoms in uranium nitride [11]. The most recent approaches involve the joint use of INS and DINS data to convey information on the anisotropy and anharmonic nature of light particle motions in condensed systems [3, 12–14]. The two techniques provide complementary information on the quantum nature of the motion on molecular systems such as water, which is characterised by proton motions reflecting the influence of internal stretching,

Journal of Physics: Conference Series **571** (2014) 012007 doi:10.1088/1742-6596/571/1/012007

bending and external librational and translational modes; these are affected by different degrees of quantization and anharmonicities [15] and the measurements of both individual energy level differences and their zero-point values allows to determine the extent of nuclear quantum effects in the system. Indeed on one hand DINS measures almost directly the momentum distribution, $n(p)$, and kinetic energy, $\langle E_K \rangle$, of the equilibrium state, that means the kinetic energy of ground and excited states for the low energy vibrations (comparable with k_BT) and, at room temperatures and below, only the ground state for the intramolecular bending and stretching modes of water molecule (because at T$\sim$300 K these modes are not excited at all). On the other hand INS in energy loss configuration measures the transition from the ground state to the first excited state, and the data are often interpreted in the frame of harmonic normal mode analysis, deriving the hydrogen-projected density of states, $g(E)_H$. The latter quantity can be used to obtain an independent estimate of the mean kinetic energy $\langle E_K \rangle$.

In cases where the system is anharmonic the values of kinetic energies derived from INS will be in general different compared to the values from DINS experiments, even at low temperatures [14]. This finding is in line with recent theoretical quantum Monte Carlo simulation studies of momentum distribution and vibrational dynamics in ice, where it was shown that the analysis of proton displacements in the harmonic assumption leads proton kinetic energy values 10% larger than those obtained directly from the Path Integral Car-Parrinello Molecular Dynamics simulation [16].

Within this framework it appears that the development of instrumentation with the capability to operate as DINS, as well as INS, spectrometers will provide a step forward in the elucidation of nuclear quantum effects [11], besides expanding the available kinematic ranges for other classes of investigations, such as the probing of electronic structures of correlated electron systems [17]. This approach has been proposed previously [18], and can be regarded as part of the long history of experiments with eV neutrons in neutron physics [19, 20]. In 1984 Brugger and co-workers [21] reported calculations and experimental tests of a prototype time of flight spectrometer for inelastic scattering using eV neutrons operating in both direct and inverse geometries. The spectrometer made use of resonant filters in the eV range to define the incident (direct geometry) or scattered (inverse geometry) neutron energies. It was reported that both geometry configurations showed energy resolutions of the same order of magnitude if the neutron flight paths are optimised. Differences in the kinematics cuts in the wave vector and energy transfers space showed that the obvious advantage of the inverse geometry is that there is no upper bound to the energy transfers which may be achieved; whereas in the direct geometry method, energy transfers are bounded by the incident energy [21]. The research programme on DINS studies for condensed matter has been mainly carried out at the ISIS facility on the eVS, then VESUVIO, spectrometer [22, 23], with early developments at the Argonne and Los Alamos National Laboratories [24, 25] and at the KENS source in Japan [26], along with more recent developments at the Bariloche neutron source (AR) [27]. All of the above instruments were built in the inverse geometry configuration, thanks to the more extended kinematic range available and for the superior energy resolution around the recoil excitation energies, as compared to direct geometry [23]. Interest in the potential of direct geometry instruments has been recently revised to extend the incident energy range up to 150 eV for the study of proton scattering cross sections [28], and, in the 3 eV range, for the probing of electronic excitations [17].

In this work we have used the SEQUOIA Fine-Resolution Fermi Chopper Spectrometer at the Spallation Neutron Source, Oak Ridge National Laboratory (U. S. A.), to carry out DINS measurements on samples of bulk water and ice at 271 K and 293 K, respectively, using incident neutrons of energy up to 6 eV. Section 2 describes the experiment and in Section 3 we discuss the data analysis and results. Section 4 reports the conclusions.

Journal of Physics: Conference Series **571** (2014) 012007 doi:10.1088/1742-6596/571/1/012007

2. Experiment

The DINS experiment was carried out on ice and liquid water at 271 K and 293 K, respectively, using SEQUOIA [29]. An high neutron incident energy, E_i= 6000 meV, was chosen in order to simultaneously access high wave vectors ($q \gtrsim 20$ Å^{-1}) and energy ($\hbar\omega \gtrsim 1000$ meV) transfers. The sample container was a flat aluminum holder, with 50 x 50 mm^2 cross-section and internal thickness 0.25 mm, with Teflon coated inner surface. The container was located perpendicular to the incident neutron beam. Data for the T= 293 K sample were recorded for approximately 3 hours, corresponding to 1.2 $\times 10^{13}$pC of proton charge collected on the spallation target, and for the T=271 K for approximately 1.5 hours, corresponding to 0.6 $\times 10^{13}$pC of proton charge collected on the spallation target. The measurements were performed for a sample in the container, the empty container, the background and a vanadium plate of the same sample geometry. Neutron counting was carried out using ^{3}He neutron detectors, covering a scattering angular range $4° < 2\theta < 56°$, with an angular increment of 0.26°. The background from the empty container was subtracted from the raw data counts, and the vanadium run provided data for detector calibration and intensity normalization of the spectra. The data treatment also included the corrections for efficiency of detectors as a function of energy and for the kinematical factor k_f/k_i.

3. Results and discussion

In Figure 1 the intensity map of the spectrum of ice obtained with nominal incident energy of E_i= 5617 meV is plotted. The horizontal axis is wave vector transfer and the vertical axis is energy transfer. We have superimposed three lines representing the expected proton recoil excitation with a recoil width equal to $\sigma_H = 5$Å^{-1}, corresponding to $\hbar\omega_R = \hbar^2 q^2/(2M)$, $\hbar\omega = [\hbar^2 q^2/(2M) \pm \hbar^2 q \sigma_H/M]$, respectively, where M is the proton mass . One can clearly identify that the inelastic scattering intensity is centred at the proton recoil excitation line. The intensity at high wave vector and low energy values originates from the zero-energy transfer elastic contribution which is broadened and dominated by the instrumental resolution, and by the oxygen recoil excitation line. It is not possible to separate these two contribution for the present value of E_i.

The intensity at the energy transfers well above the proton recoil width is due to a background from fast delayed neutrons partially overlapping with neutrons selected with Fermi chopper. This contribution is independent of the scattering angle and is shown in Figure 2 for the ice sample. In this figure the fixed angle scattering intensities are reported for a series of scattering angles between 4 degrees and 56 degrees, and show that the background is the same at all scattering angles and is non negligible for energy transfers above 4500 meV. This finding is confirmed by inspection of the time of flight spectrum recorded by the SEQUOIA incident neutron monitor, Monitor 1; the spectrum reported in Figure 3 is acquired by a phasing and opening of the background T0 chopper just before the requested E_i=5617 meV, and it is shown that neutrons selected with the Fermi chopper with energy 5617 meV are superimposed on the tail of fast delayed neutrons.

We found that the background on the fixed angle spectra can be modeled by a fifth-order polynomial.

The analysis of DINS data is carried out within the framework of the Impulse Approximation (IA) [22], where the incident neutron wavelengths are much smaller than the inter-atomic spacing and thus atoms scatter incoherently, with scattered intensity being the sum of intensities from individual atoms in the sample, with conservation of the total kinetic energy and momentum of the neutron and the atom. In the IA regime the inelastic neutron scattering cross section is related in a simple way to the atomic momentum distribution, $n(\mathbf{p})$. The neutron scattering

Journal of Physics: Conference Series **571** (2014) 012007 doi:10.1088/1742-6596/571/1/012007

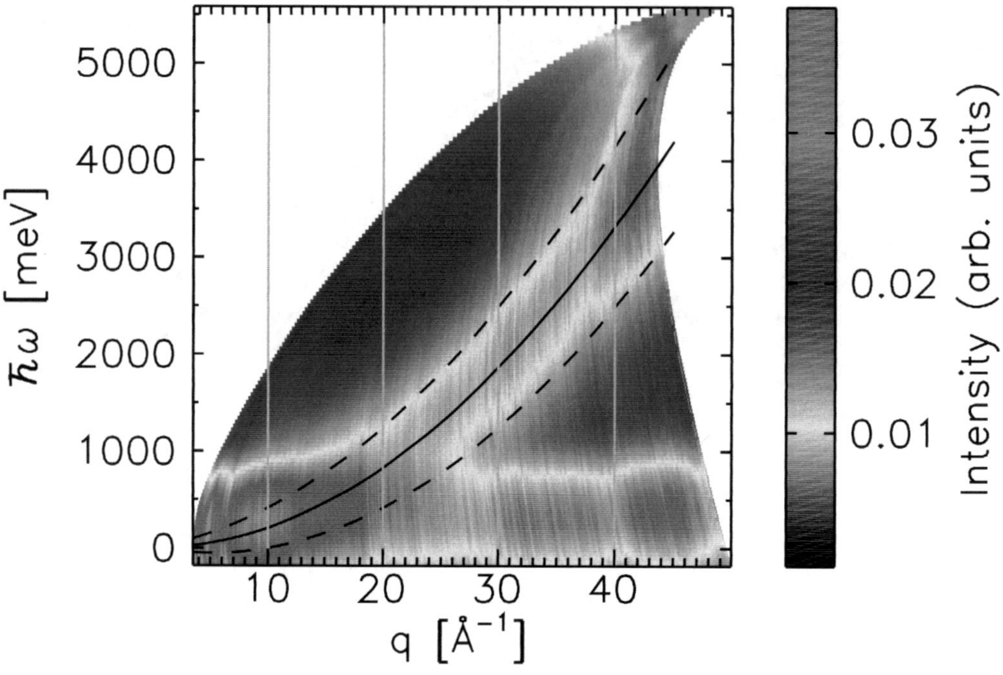

Figure 1. (Color online) Scattering intensity map of energy-wave vector space obtained on SEQUOIA using incident neutrons of 6000 meV (effective energy of 5617 meV) at T=271K. The theoretical proton recoil excitation lines with a width of 5 Å^{-1} are reported as continuous and dashed black lines.

function $S_{\text{IA}}(\mathbf{q}, \omega)$ is:

$$S_{\text{IA}}(\mathbf{q}, \omega) = \int \mathbf{n}(\mathbf{p})\, \delta \left(\omega - \omega_{\mathbf{R}} - \frac{\mathbf{p} \cdot \hat{\mathbf{q}}}{\mathbf{M}} \right) \mathrm{d}\mathbf{p} \tag{1}$$

$$\frac{\hbar q}{M} S_{\text{IA}}(\mathbf{q}, \omega) = J_{\text{IA}}(y, \hat{\mathbf{q}}) = \int n(\mathbf{p}) \delta \left(y - \mathbf{p} \cdot \hat{\mathbf{q}} \right) \mathrm{d}\mathbf{p} \tag{2}$$

where $(\mathbf{q}, \omega)$ are the wave vector and energy transfers, M is the mass of the struck atom , $y = \frac{M}{\hbar q} \left[\omega - \frac{\hbar q^2}{2M} \right]$ is the particle momentum along the $\hat{\mathbf{q}}$ direction, and $J_{\text{IA}}(y, \hat{\mathbf{q}})$ is the neutron Compton Profile (NCP) [22]. [1] The above equations assume that the measurements are carried out at constant wave vector transfer and varying only the energy. In practice, the width and intensity of the scattering signal depend on the trajectory in the $\hbar\omega, q$ plane of the measurement (see Fig. 1). The trajectories are different in the case of inverse and direct geometry [23, 28, 30]. In particular, on VESUVIO, the single detector trajectories intersect the proton recoil line more tangentially [23] than on SEQUOIA. As shown in Equation 2, the width of the scattering arises from the width of the proton recoil line shape (see Fig. 1), which is dependent on the width of the proton momentum distribution in the system. Indeed, a reliable DINS experiment should include the measurement of the full recoil line shape, *i.e.* extending for at least four standard

[1] For consistency with previous literature and ease of notation we write the momentum as a wave vector.

Journal of Physics: Conference Series **571** (2014) 012007 doi:10.1088/1742-6596/571/1/012007

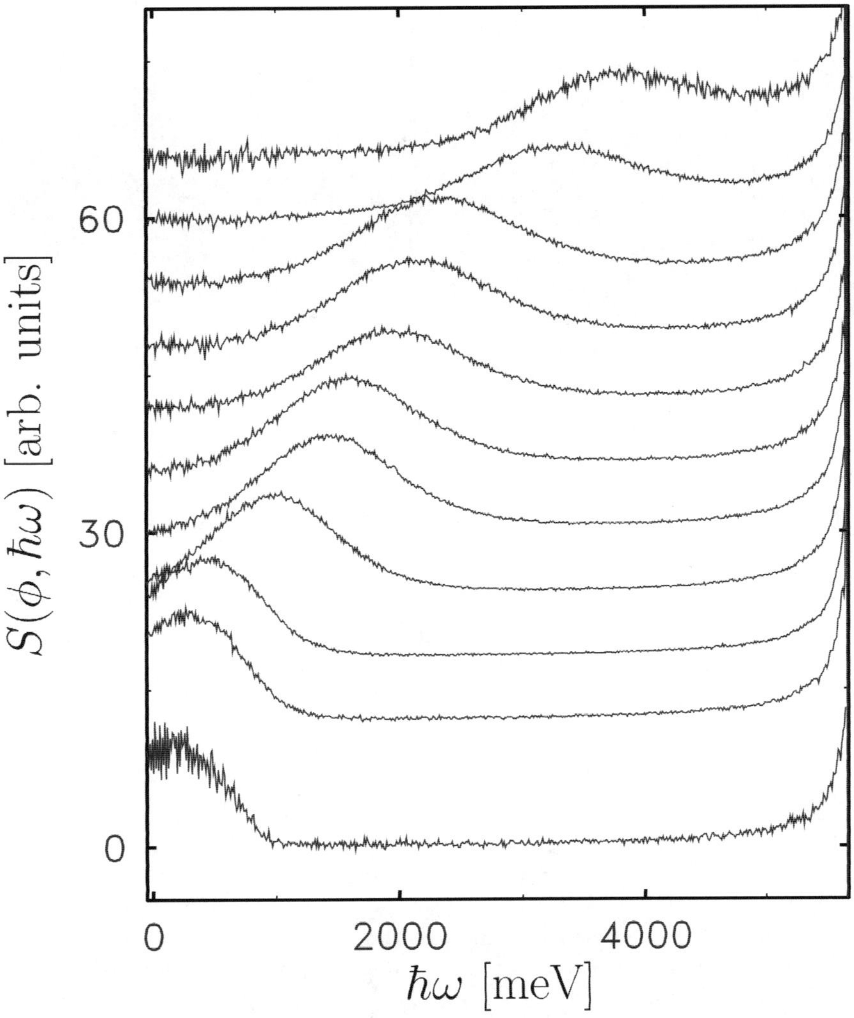

Figure 2. (Color online) Fixed angle scattering intensities for a selection of scattering angles between 4 degrees and 56 degrees; these were obtained using standard routines available on the instrument, and are shifted upwards for clarity.

deviations from the centroid of the recoil peak [31]. Typically, the standard deviations of the proton recoil peaks are of the order of $\hbar^2 q\sigma_H/M$, where σ_H, being of the order of 5 Å^{-1}, is the standard deviation of the momentum distribution, $n(\mathbf{p})$. Figure 4 reports an example of

Journal of Physics: Conference Series **571** (2014) 012007 doi:10.1088/1742-6596/571/1/012007

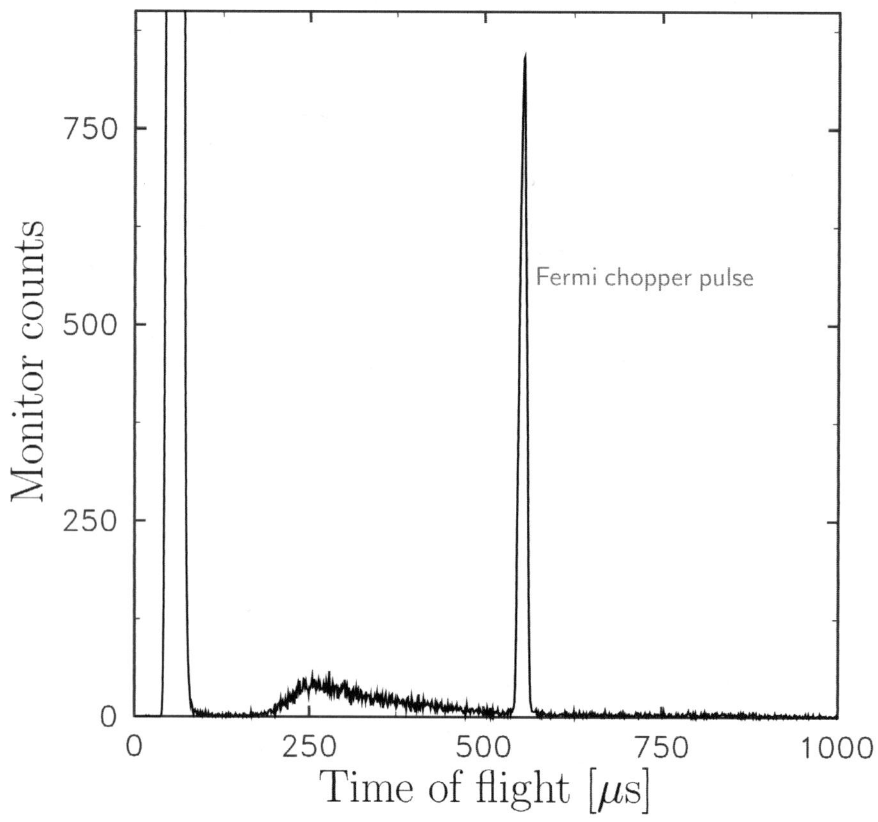

Figure 3. (Color online) Example of time of flight intensity recorded by the incident Monitor 1, placed at 18.23 m from the moderator. The pulse from the Fermi chopper is centred at approximately 550 μs . The peak at approximately 250 μs is a consequence of monitor dead-time after saturation of the monitor itself at short time, due to very large neutron flux (of high energy neutrons), and transmission of the delayed fast neutrons through both the background T0 and Fermi choppers.

the crossings of the detector trajectories and the proton recoil at different widths of the recoil line shape. The central recoil line marks the energy and wave vector positions of the centroid of the scattered intensity, which is crossed once for each detector trajectory. The recoil lines corresponding to a broadening of one standard deviation (red dot-dashed lines) cross the detector trajectories twice, therefore a fixed angle scan samples the same proton momentum value two times (red crosses in Fig. 4). The same behaviour is found for two and three(blue short-dashed lines) -standard deviations broadening.

Indeed the measured value of proton momentum, y, depends on the distance between the detector trajectory and the recoil trajectory, that is on the difference $\frac{M}{\hbar q}\left[\omega - \frac{\hbar q^2}{2M}\right]$, which is non-monotonic (see Fig. 4) and causing double counting of the scattering intensity for $y \geq 5\text{Å}^{-1}$.

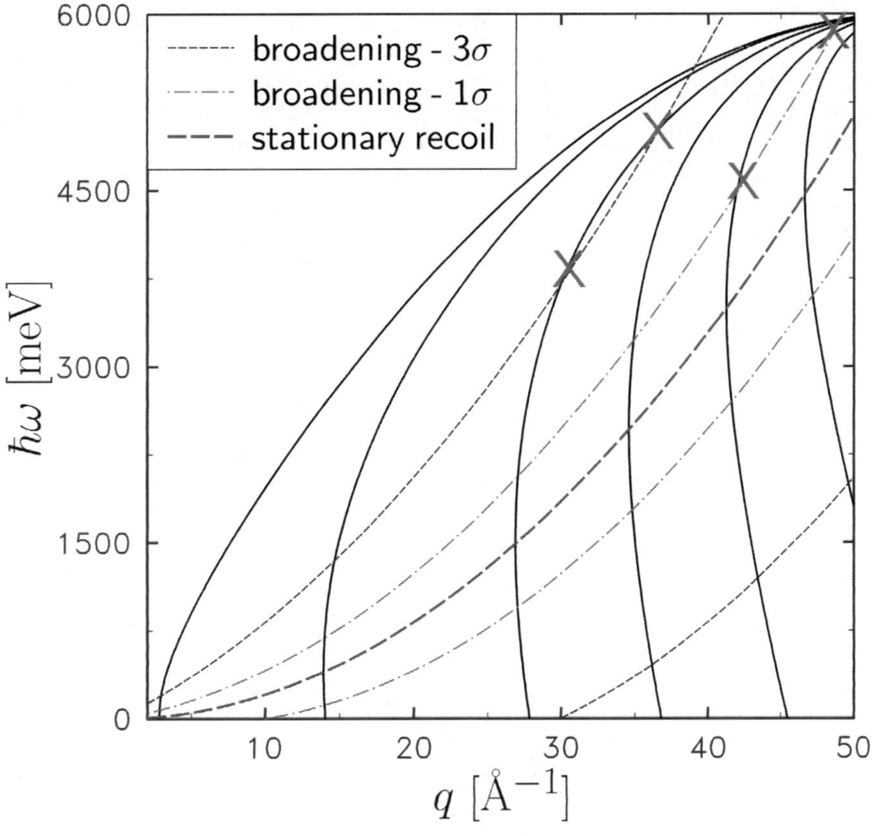

Figure 4. (Color online) Crossing of detector trajectories (black continuous lines), for representative scattering angles between 4 degrees and 60 degrees, with the proton recoil line (green dashed line). The red dot-dashed lines and blue short-dashed lines represent the broadening of the recoil due to finite momentum width corresponding to one (red dot-dashed lines) and three (blue short-dashed lines) standard deviations. The X signs are examples of the intersections between the detector trajectories and the momentum-broadened proton recoil.

This means that a line shape analysis of the spectra measured at constant scattering angle cannot provide a direct information on the shape and intensity of $S_{\text{IA}}(\mathbf{q}, \omega)$, due to the double counting of the same proton momentum at different $\hbar\omega, q$ values. It has been proposed that the use of a Waller-Froman Jacobian relating the free recoil excitation and the wave vector transfer as a function of the scattering angle can be employed to relate a constant-angle scan to a constant-q scan [28]. Indeed the above Jacobian is calculated assuming $\hbar\omega \equiv \hbar^2 q^2/(2M)$, that is, a non-physical situation in which all proton atoms have zero momentum, which is never realised in nature. The resulting Jacobian-transformed spectra would contain an overestimation of the high proton momentum tail of $S_{\rm IA}(\mathbf{q}, \omega)$ [28]. Based on the above findings we conclude that DINS measurements on direct geometry instruments cannot be reliably analysed using constant scattering angle spectra. On the other hand, constant-q slices allow to correctly analyse the DINS spectra.

Data for ice, water, and the corresponding empty container signals, have then been transformed into constant-q slices for three q values, namely $q = 30 \pm 1$ Å^{-1}, $q = 34 \pm 1$ Å^{-1}, $q = 40 \pm 1$ Å^{-1}. These values allow to obtain recoil spectra extending to proton momentum values $\mathbf{p}\cdot\hat{\mathbf{q}} \simeq \pm$ 20 Å^{-1}. In this respect there are specific advantages and disadvantages of direct geometry compared to inverse geometry DINS measurements for the proton: the advantage is that in principle the fine detector coverage and the kinematic plane of the direct geometry allow an extended choice of q slices; the disadvantage is that the $\mathbf{p}\cdot\hat{\mathbf{q}}$ range accessed is $\simeq$ 35 % smaller than on VESUVIO ($\mathbf{p}\cdot\hat{\mathbf{q}} \simeq \pm$ 20 Å^{-1} compared to $\mathbf{p}\cdot\hat{\mathbf{q}} \simeq \pm$ 30 Å^{-1} for inverse geometry) because of the kinematic constraints (see , Fig.1 and Fig.4), and this limits the choice of suitable constant-q slices at high wave vector transfers, as shown in Fig.4 . The three spectra for the ice sample measurement are reported in Figure 5. The figure shows the shift of the recoil centroids to higher energy transfers at increasing values of wave vector transfer, as expected from the Impulse Approximation prediction. The high energy transfers tail shows the contribution due to background signals from fast neutrons as discussed above (see Fig. 2) . The spectra have been transformed in momentum space, rebinned to constant bin width of 0.5 Å^{-1} and normalised, yielding the neutron Compton profiles for the samples investigated. It has to be stressed that Equation 2 is assumed to be strictly valid in the limit of infinite wave vector transfer. From the experimental point of view, the deviations from the asymptotic regime, the broadening due to the instrumental resolution, and the isotropic nature of the sample, are accounted for by the following representation of the NCP [22]:

$$F(y, q) = [J_{\rm IA}(y) + \Delta J(y, q)] \otimes R(y, q). \tag{3}$$

where $\Delta J(y, q)$ accounts for the finite-q correction terms (final state effects) [32], $R(y, q)$ is the instrumental resolution function and for an isotropic sample the particle momentum distribution is spherically averaged and depends only on the modulus of $\mathbf{p}$, therefore $J_{\rm IA}(y) = 2\pi \int_{|y|}^{\infty} p\; n(p)\; dp$ [22].

Figure 6 reports the three experimental NCP for the ice and water samples, respectively. The plots show the raw differences, $[F(y, q)_{ice} - F(y, q)_{water}]$, between ice and water ; these are directly related to the differences of proton momentum distributions in the two samples. The differences are small, but it can be noticed that $F(y, q)_{ice}$ has a slightly lower height at the peak centre and larger wings in the ranges -10 Å$^{-1} \leq y \leq -5$ Å^{-1}, 5 Å$^{-1} \leq y \leq 10$ Å^{-1}. This is indeed shown by the differences which are positive, on average, in the above ranges, and negative around the peak centre. Such a behaviour suggests that the neutron Compton profile, and thus the proton momentum distribution, in ice at 271 K is broader than liquid water at room temperature. Figure 7 reports a comparison between the $F(y, q)$ for ice for $q = 34 \pm 1$ Å^{-1} from the present experiment and the corresponding angle averaged- NCP, $\bar{F}(y)$, from a previous VESUVIO measurement on the same sample at the same temperature, T=271 K [2]. It can be noticed that the shapes are similar, with the exception of a reduced momentum range for the SEQUOIA measurement and the presence of the above mentioned background.

In order to analyse the line shapes of the NCP for the present measurements, the data have been fitted using the ice NCP as a reference: the model line shape for $F(y, q)$ at the three q values is a Compton profile of the form $[J_{\rm IA}(y) + \Delta J(y, q)]$ with parameters fixed to the values from ref. [2], broadened by the resolution function, $R(y, q)$; the latter is assumed to be a

Journal of Physics: Conference Series **571** (2014) 012007 doi:10.1088/1742-6596/571/1/012007

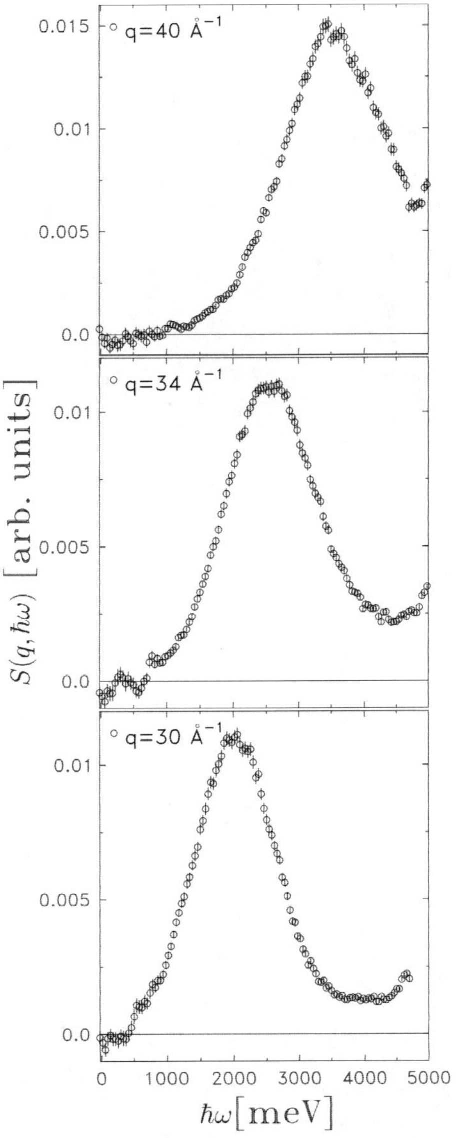

Figure 5. (Color online) Proton recoil spectra for the ice sample at $q = 30 \pm 1$ Å^{-1} (lower panel), $q = 34 \pm 1$ Å^{-1} (middle panel), $q = 40 \pm 1$ Å^{-1} (top panel)

Gaussian with the standard deviation as a free parameter, plus a parabolic function to model the background at high momenta. The final state effects term, $\Delta J(y,q)$, is expected to be smaller on VESUVIO data than on SEQUOIA, since its magnitude is proportional to inverse powers of q, being the q-range on VESUVIO extended to values one order of magnitude larger than on SEQUOIA [33]. However, preliminary tests on the sensitivity of the line shape modeling showed that, for the present data, the contribution of final state effects terms of order $1/q^2$ and beyond were negligible. The model function was fitted to the data using a nonlinear minimization

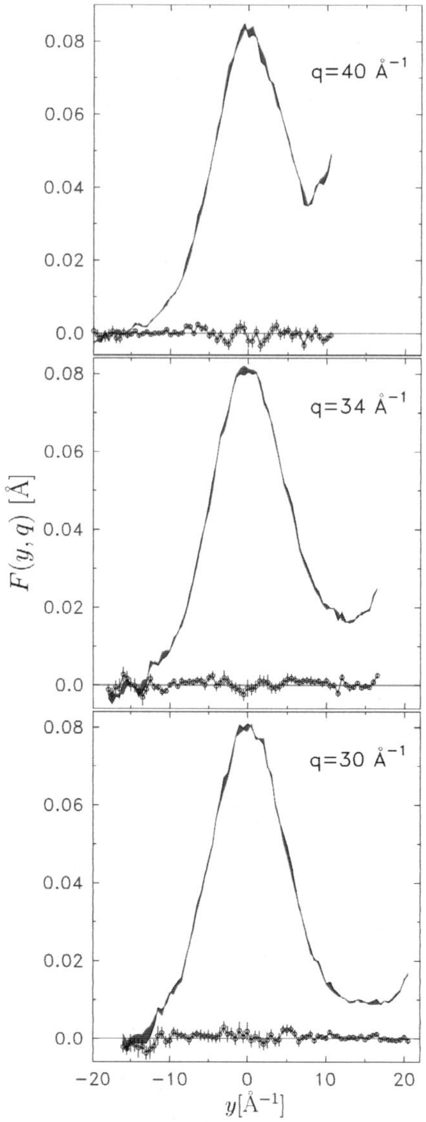

Figure 6. (Color online) Proton neutron Compton profiles for the ice (black lines) and water (red lines) samples at $q = 30 \pm 1$ Å^{-1} (lower panel), $q = 34 \pm 1$ Å^{-1} (middle panel), $q = 40 \pm 1$ Å^{-1} (top panel). The green filling between the profiles at each q value highlights the differences between ice and water. The panels also report the raw differences, $[F(y,q)_{ice} - F(y,q)_{water}]$ as black dots with error bars, which oscillate around zero.

routine described in ref. [2]. The fitting provided the three sets of parameters describing the resolutions and background shapes at the three q values.

Subsequently, the fitting procedure has been applied to the water data by fixing the resolution and background parameters to the values found for ice, while the width (standard deviation)

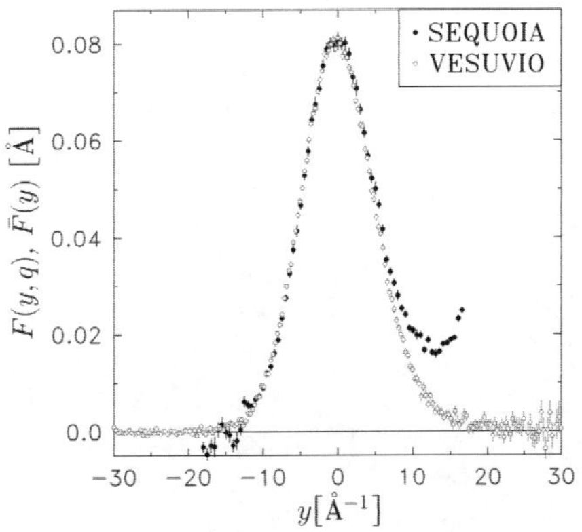

Figure 7. (Color online) Proton neutron Compton profiles for ice (solid black circles with error bars) from the present measurement, and the corresponding angle averaged- NCP, $\bar{F}(y)$, from a previous VESUVIO measurement on the same sample at the same temperature, T=271 K (red open circles) [2].

of $J_{\rm IA}(y)$ for water is a free fitting parameter. The results of the analysis of the water data provided an estimate of the standard deviation of the proton momentum distribution in water, averaged over the three q values, namely: $\sigma_{water} = (4.91 \pm 0.03)$ Å^{-1}. This value corresponds to a proton kinetic energy, $\langle E_K \rangle_{water} = \frac{3\hbar^2}{2M}\sigma^2_{water} = 151 \pm 2$ meV. This has to be compared with the proton kinetic energy in ice at T=271 K, namely 156 ± 2 meV from ref. [2]. This finding confirms qualitatively previous comparisons of the experimental determinations of proton kinetic energy in ice close to the triple point and liquid water at room temperature [34]. Theoretical works on the determination of proton kinetic energies in water relate the decrease in hydrogen bonding in the liquid, and the more flexible environment, to be associated to either a negligible difference between the kinetic energies in the solid and the liquid, or to a nearly continuous behavior [12]. The present result suggest on the other hand an unexpected decrease of in the liquid phase, which should be investigated with higher accuracy measurements and calculations. Interestingly, the deuterium kinetic energy in heavy water does not show such a behavior, being the kinetic energy in the solid close to the triple point slightly lower than in the room temperature liquid, in agreement with theory [3]. The present results suggests that the different degrees of anharmonicity and hydrogen bonding in light water compared to heavy water are at the origin of these quantum differences in the local environment of the proton in the solid and liquid phases.

4. Conclusions

The measurements presented here allowed an assessment of the performance of SEQUOIA for DINS measurements of proton momentum distribution. The data analysis confirmed qualitatively that the kinetic energy in room temperature water is lower than in ice close to the triple point, in line with previous VESUVIO measurements. It is suggested that this property is ascribed to the competition between hydrogen bonding and anharmonicity in the two phases, with different influence of quantum effects with respect to heavy water. From the instrumental point of view, the complementarity between direct and inverse-types of instrument can convey information on specific advantages and disadvantages of chopper spectrometer with respect to inverse geometry spectrometers, namely: 1) Count rate and flexibility; at present SEQUOIA has a factor of $\simeq 8$ larger counting capabilities over VESUVIO for this type of measurements. This is mostly due to the higher neutron source flux with respect to ISIS. The flexibility allows to operate with a large number of incident energies, while on VESUVIO this choice is limited to one final energy value. 2) Detector solid angle coverage; this allows to operate constant-q slices for data analysis with a larger flexibility compared to VESUVIO [33]. 3) Kinematic trajectories; these are limited with respect to VESUVIO and do not allow for a complete scan of the full proton momentum range, with a reduction of $\simeq 35$ % with respect to VESUVIO. The kinematic trajectories do not allow to analyse constant-angle spectra. More optimised data sets in this respect may be obtained for studies on deuteron containing samples [35]. Measurements on even heavier masses would however lack the mass-specific separation capability [36, 37]that can be achieved on VESUVIO [9]. 4) Fast neutrons background and resolution; the optimised phasing in the current set up does not allow to completely suppress the background, preventing fine line shape analysis beyond the determination of the standard deviation. Such optimisation should require the use of sequential choppers. The proton momentum space resolution is presently worse than on VESUVIO [23,38], limiting the line shape analysis capability.

The above considerations show that direct geometry instruments can be used for large energy transfers to provide complementary information to inverse geometry instruments, with a series of limitations that still do not allow to reach the capabilities of inverse geometry configuration. Substantial improvements on the instrumental resolution can be achieved with new chopper design to optimise slits curvature and distances. These optimisation strategies would be however most effective in the case of deuteron studies rather than for proton ones.

5. Acknowledgments

This work was supported within the with the support of META (Materials Enhancement for Technological Applications) PIRSES-GA-2010-269182, Marie Curie Actions, People, VII Framework Programme, and within the CNR-STFC Agreement No. 06/20018 concerning collaboration in scientific research at the spallation neutron source ISIS. The neutron scattering experiment at Oak Ridge National Laboratorys Spallation Neutron Source was sponsored by the Scientific User Facilities Division, Office of Basic Energy Sciences, US Department of Energy.

References

[1] Reiter G F, Senesi R and Mayers J 2010 *Physical Review Letters* **105** 148101 (*Preprint* 1002.4398)
[2] Flammini D, Pietropaolo A, Senesi R, Andreani C, McBride F, Hodgson A, Adams M A, Lin L and Car R 2012 *The Journal of Chemical Physics* **136** 024504
[3] Romanelli G, Ceriotti M, Manolopoulos D E, Pantalei C, Senesi R and Andreani C 2013 *The Journal of Physical Chemistry Letters* **4** 3251–3256 (*Preprint* http://pubs.acs.org/doi/pdf/10.1021/jz401538r) URL http://pubs.acs.org/doi/abs/10.1021/jz401538r
[4] Krzystyniak M and Fernandez-Alonso F 2011 *Physical Review B - Condensed Matter and Materials Physics* **83** 134305
[5] Krzystyniak M and Abdul-Redah T 2010 *Physical Review B - Condensed Matter and Materials Physics* **82** 064301
[6] Krzystyniak M 2010 *Journal of Chemical Physics* **133** 144505
[7] Krzystyniak M, Adams M A, Lovell A, Skipper N T, Bennington S M, Mayers J and Fernandez-Alonso F 2011 *Faraday Discuss.* **151**(0) 171–197 URL http://dx.doi.org/10.1039/C1FD00036E
[8] Seel A G, Ceriotti M, Edwards P P and Mayers J 2012 *Journal of Physics Condensed Matter* **24** J5401
[9] Krzystyniak M, Richards S, Seel A and Fernandez-Alonso F 2013 *Physical Review B - Condensed Matter and Materials Physics* **88** 184304
[10] Price D and Fernandez-Alonso F 2013 *Experimental Methods in the Physical Sciences* **44** 1–136
[11] Aczel A A, Granroth G E, MacDougall G J, Buyers W J L, Abernathy D L, Samolyuk G D, Stocks G M and Nagler S E 2012 *Nature Communications* **3** 1124 (*Preprint* 1206.3597)
[12] Finkelstein Y and Moreh R 2014 *Chemical Physics* **431** 58–63
[13] Andreani C, Romanelli G and Senesi R 2013 *Chemical Physics* **427** 106–110

Journal of Physics: Conference Series **571** (2014) 012007 doi:10.1088/1742-6596/571/1/012007

[14] Senesi R, Flammini D, Kolesnikov A I, Murray É D, Galli G and Andreani C 2013 *Journal of Chemical Physics* **139** 074504
[15] Marechal Y 2006 *The Hydrogen Bond and the Water Molecule: The Physics and Chemistry of Water, Aqueous and Bio-Media* (Elsevier Science) ISBN 9780080469294
[16] Lin L, Morrone J A, Car R and Parrinello M 2011 *Physical Review B* **83** 220302
[17] Kim Y J, Sorini A P, Stock C, Perring T G, van den Brink J and Devereaux T P 2011 *Physical Review B* **84**(8) 085132 URL `http://link.aps.org/doi/10.1103/PhysRevB.84.085132`
[18] Silver R N 1984 *Workshop held in Los Alamos, N. Mex., 13-15 Feb. 1984* pp 345–648
[19] Lamb W E 1939 *Physical Review* **55** 190–197
[20] Gol'danskii V I 1957 *Soviet Phys. JETP* **4** 604
[21] Brugger R M, Taylor A D, Olsen C E, Goldstone J A and Soper A K 1984 *Nuclear Instruments and Methods in Physics Research* **221** 393 – 407 ISSN 0167-5087 URL `http://www.sciencedirect.com/science/article/B73DP-478HFR2-20/2/c56fd981fa54c58efa2c6d5177da312c`
[22] Andreani C, Colognesi D, Mayers J, Reiter G F and Senesi R 2005 *Advances in Physics* **54** 377
[23] Pietropaolo A and Senesi R 2011 *Physics Reports* **508** 45–90
[24] Crawford R K 1986 *Proceedings of the Ninth International Collaboration on Advanced Neutron Sources* pp 397–430
[25] Taylor A, Robinson R A and Seeger P A 1984 *Nuclear Instruments and Methods in Physics Research* **224** 133–141
[26] Carpenter J and Watanabe N 1983 *Nuclear Instruments and Methods in Physics Research* **213** 311 – 316 ISSN 0167-5087
[27] Rodríguez Palomino L A, Blostein J J and Dawidowski J 2011 *Nuclear Instruments and Methods in Physics Research A* **646** 142–152
[28] Stock C, Cowley R A, Taylor J W and Bennington S M 2010 *Physical Review B* **81** 024303 (*Preprint* `0907.1945`)
[29] Granroth G E, Kolesnikov A I, Sherline T E, Clancy J P, Ross K A, Ruff J P C, Gaulin B D and Nagler S E 2010 *Journal of Physics: Conference Series* **251** 012058
[30] Brugger R 1984 *Nuclear Instruments and Methods in Physics Research* **221** 393–407
[31] Senesi R 2012 *Nuclear Instruments and Methods in Physics Research A* **661** 70–76
[32] Sears V F 1984 *Phys. Rev. B* **30** 44–51
[33] Senesi R, Pietropaolo A and Andreani C 2008 *Nuclear Instruments and Methods in Physics Research A* **594** 244–252
[34] Senesi R, Romanelli G, Adams M and Andreani C 2013 *Chemical Physics* **427** 111–116
[35] Diallo S O private communication
[36] Seel A, Krzystyniak M and Fernandez-Alonso F *Journal of Physics: Conference Series* This issue
[37] Krzystyniak M, Seel A, Richards S, Gutmann M J and Fernandez-Alonso F *Journal of Physics: Conference Series* This issue
[38] Mayers J, Gidopoulos N I, Adams M A, Reiter G, Andreani C and Senesi R 2011 *Phys. Rev. B* **84**(5) 056301

Journal of Physics: Conference Series **571** (2014) 012008 doi:10.1088/1742-6596/571/1/012008

New prospects of VESUVIO applied to measurements in water mixtures

L. A. Rodríguez Palomino[1], J. Dawidowski[1, 3], J. J. Blostein[1], G. J. Cuello[2]

[1] Consejo Nacional de Investigaciones Científicas y Técnicas (CONICET), Comisión Nacional de Energía Atómica-Universidad Nacional de Cuyo, Argentina
[2] Institut Laue Langevin, 71, Av. des Martyrs, 38042 Grenoble, France

E-mail: `javier@cab.cnea.gov.ar`

Abstract. We present new measurements on mixtures of light and heavy water in the spectrometer VESUVIO (Rutherford Appleton Laboratory, UK), and analyze them from the perspective of different kind of applications. We perform a single detector analysis and show the multiple scattering and attenuation corrections with the aim to employ them in mass-spectrometry. We also show the capabilities to perform transmission measurements to determine total cross sections of an acceptable quality by means of its transmission monitor.

1. Introduction

In this paper we discuss the results of new measurements of various mixtures of light water and heavy water at spectrometer VESUVIO (ISIS, Rutherford Appleton Laboratory, UK). The use of water as a system of interest, far from being considered as a new chapter in an old debate, is used here as an experience to highlight different capabilities of VESUVIO that had not been employed intensively so far. Although designed as an instrument to explore momentum distributions, VESUVIO provides information on scattered neutrons on a wide range of angles and energies. In fact it is one of the few available instruments that offers the possibility of reaching epithermal energies up to about 100 eV (exclusive property of accelerator-based sources), while having available a substantial flux of thermal neutrons [1]. Also, it has a transmission monitor, placed at a distance from the source that makes it capable to measure energy-dependent transmission spectra with very good resolution in energy. Given these possibilities, the feasibility of finding new uses for this facility has already been begun to be explored. Such is the case of Mass-selective neutron spectroscopy recently explored by Krzystyniak *et al.* [2].

In this work we focus out attention on electron-Volt spectroscopy applications that do not need to go through transformation from time-of-flight to the momentum transfer y-scale [3]. In those cases it is more convenient to work on individual detector signals in the time-of flight scale [4]. We select the detectors that show the best signals and show the steps necessary to perform Monte Carlo simulations to calculate multiple scattering and attenuation corrections, a prerequisite aiming to employ single-detector information in mass spectrometry. In the description of the procedure, we summarize the steps of the analysis, that comprise the

[3] Corresponding author

Journal of Physics: Conference Series **571** (2014) 012008 doi:10.1088/1742-6596/571/1/012008

measurement of the incident neutron spectrum, and the definition of the detectors efficiency function, necessary to perform the Monte Carlo simulations to correct by multiple scattering and sample attenuation effects, which are briefly addressed.

We also pay particular attention to the use of the transmission monitor of VESUVIO to determine total cross sections, a feature which to the author's knowledge has not been exploited to date. For this purpose, we examine the total cross sections of different H_2O/D_2O mixtures measured at VESUVIO compared with measurements performed at the Bariloche Electron LINAC, which has been employed in this field for decades [5]. We show that good quality total cross sections can be measured in several decades of energy.

2. Experimental

2.1. Samples

The measurements were performed on three different water mixtures at room temperature. The deuterium molar concentrations of the mixtures were $x_D = 0$ (light water), 0.25, and 0.5. A coin-shaped aluminum can 1.4 mm thick and 5.01 cm diameter was employed as a container for all the samples. The sample covered the entire penumbra of the beam at the sample position, necessary condition to perform both transmission and standard Deep Inelastic Neutron Scattering (DINS) experiments at VESUVIO. The same mixtures (prepared at ISIS) were employed at the Bariloche Electron LINAC to perform transmission measurements.

2.2. VESUVIO setup

We will follow the nomenclature employed in Ref. [4], where a complete description of VESUVIO is found. YAP γ-ray detectors (S135 to S198) placed in forward scattering direction [6] were employed in the analysis of the present work. Detector S2 is a ^{6}Li doped glass detector placed at 13.43 m from the neutron moderator was employed as transmission monitor, while detector S1 located 8.6 m from the moderator was used as the incident beam monitor, whose number of counts was used to normalize all the measured spectra. The sample is located inside an evacuated bell, at 11 m from the moderator. In addition to the measurements on water samples mentioned above, ancillary measurements were performed. A slab of width 2 mm lead was employed as a standard sample for calibration, and also measurements of empty-cell and empty instrument were performed.

3. Basic data analysis formalism

The basic expression for the difference filter-out minus filter-in spectra in the time of flight scale is based on the expression stated by Powles [7] for scattered spectra in the time of flight scale

$$c(t,\theta)\Delta t = \int_{\substack{\mathrm{E_{inf}} \\ \mathrm{t=const}}}^{\infty} dE_0\, \Phi(E_0) \frac{d^2\sigma(E_0,E,\theta)}{d\Omega dE}\, \varepsilon(E) \left[1 - e^{-nT\sigma_F(E)}\right] \left|\frac{\partial E}{\partial t}\right| \Delta\Omega \Delta t, \quad (1)$$

where $\Phi(E_0)$ is the energy dependent incident neutron flux, $\frac{d^2\sigma(E_0,E,\theta)}{d\Omega dE}$ the double-differential cross section of the sample, $\varepsilon(E)$ the detector efficiency, $(1 - e^{-nT\sigma_F(E)})$ the absorption probability of the resonant filter, characterized by a number density n, a thickness T, and a total cross section $\sigma_F(E)$, and $\Delta\Omega$ the solid angle subtended by the detectors. The lower limit of integration is determined by kinematic condition that in the second flight path the neutron has infinite velocity, so $E_{\mathrm{inf}} = \frac{1}{2}m\frac{L_0^2}{t^2}$, where m is the neutron mass and L_0 the source-sample distance. In Reference [8] we show the link between the calculated spectrum (Eq. (1)), which is a microscopic magnitude, and that measured (comprising multiple scattering events, and container contributions). In the further sections we will show how to access to the required magnitudes employed in Eq. (1) from the experimental information taken at VESUVIO.

4. Results

In this section we present the results of two different types that can be obtained in VESUVIO. On one side we show the analysis of the data obtained from one of the selected YAP detectors in forward scattering direction, showing the multiple scattering and attenuation corrections, with the aim to evaluate its use as mass spectrometer. In addition we analyze the use of the transmission monitor signal to determine total cross sections. All the experimental data treatment we show in this work was performed with programs based on the Open Genie display and analysis of data suite, developed for the neutron scattering instruments at the ISIS facility [9].

4.1. YAP detector analysis

4.1.1. Detector selection As a first step of our analysis, we selected the detectors to work with. The selection of detectors was based on the criterion that the difference filter-out minus filter-in spectra showed a good baseline, free of the filter-dependent gamma background described by Mayers [10]. Thus, in our further work we will not apply the background calculations there described. Fig. 1 shows one of the selected spectra of H_2O by comparison with another in which the baseline shows a systematic behavior that affects the region of interest (between 0 and 500 μsec), and was thus discarded. With this criterion we selected for the analysis detectors S162 (located at a distance of 75.1 cm from the sample at a forward scattering angle of 32.76^o) and S163 (located at a distance of 75.6 cm from the sample at a forward scattering angle of 42.84^o).

4.1.2. Detector efficiency function The knowledge of the efficiency function of each detector is essential to perform the analysis based on Eq. (1). To determine it we applied a procedure similar to that described in [8], where knowledge of the incident spectrum is essential [11]. We firstly measured the incident spectrum with the transmission monitor S2 located at 13.44 m from the source, leaving empty the sample position. The measured spectra, corrected by detector efficiency as described below, is the incident spectrum $\Phi_0(E)$. We also measured the elastically scattered neutron spectrum by a lead sample $\Phi_{Pb}(E)$ at each detector, and also its background $\Phi_{Bg}(E)$ leaving empty the sample position. The efficiency can be then determined as

$$\varepsilon(E) = \frac{\Phi_{Pb}(E) - \Phi_{Bg}(E)}{\Phi_0(E)}. \tag{2}$$

As discussed in [8], this is a zero-order approach to the efficiency calculation. A more refined determination should include the subtraction of multiple scattering and inelastic effects. In this work we will omit such corrections since we proved that their contribution is secondary. Fig. 2 shows $\Phi_0(E)$, $\Phi_{Pb}(E)$ and $\Phi_{Bg}(E)$ (measured with detector S163, at a scattering angle of 42.8^o in the forward direction, located at 75.6 cm of the sample). All the spectra were normalized by monitor counts measured by transmission monitor S1, located at 8.61 m from the source. In the inset we show he efficiency of detector S163 determined from Eq. (2). Leaving aside the thermal region where Bragg peaks appear and thus lead cannot be used as a mirror, the efficiency determined at the epithermal region shows peaks due to the resonances in the gold foils that cover the detectors. In particular, we note the resonance at 4.9 eV.

4.2. Experimental and Monte Carlo results

In Figure 3 we show the experimental spectra in time-of flight scale for detectors S162 and S163 and the samples of H_2O and the water mixtures H_2O/D_2O for $x_D = 0.25$ and 0.5. Since we use single detectors for our analysis, the time of flight data were rebinned to 5 μsec intervals in order to improve statistics shown in each channel. Also in the same figure we show the Monte Carlo results commented below. For the particular case of $x_D = 0.25$ at a scattering angle of

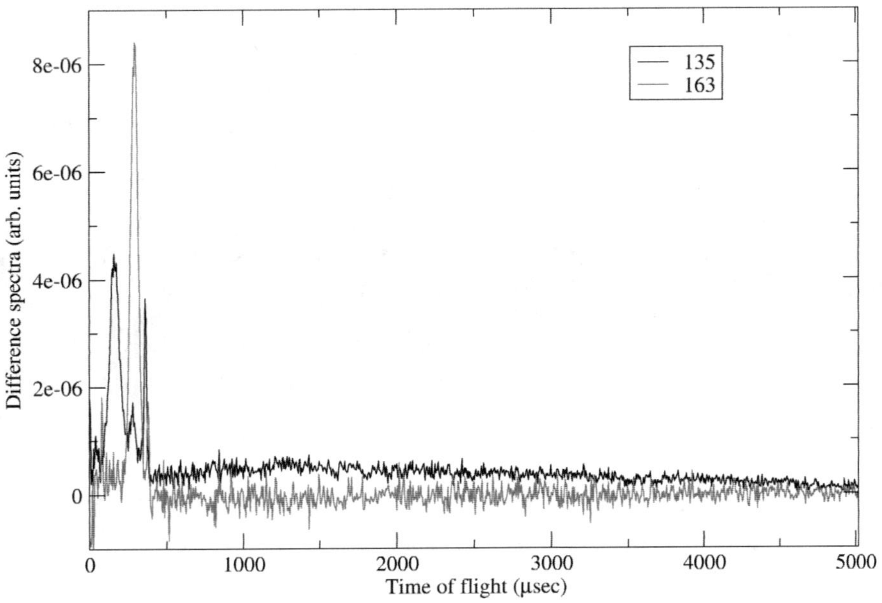

Figure 1. Comparison between two filter difference spectra (filter-out minus filter-in) in H_2O showing the stability of the baseline. Spectrum S163 shows no background and was selected for the analysis.

42.8^o we show the contributions of H (light shaded area), D (medium shaded area), and O (dark shaded area) involved in the calculation.

The basic principles on which the Monte Carlo code is based, are exhaustively described in Ref. [12], so only a brief description will be given here. For the purposes of this work, in Monte Carlo calculations the samples were modeled considering a free-atom model as in Ref. [13], with effective temperatures calculated employing Granada's Synthetic model [14]. Thus for hydrogen we considered $(k_BT)_H = 115.2$ meV, for deuterium $(k_BT)_D = 80.11$ meV and for oxygen $(k_BT)_O = 36.86$ meV. HDO molecules were not considered. They have slightly different vibrational modes that produce minor changes in those effective temperatures. Thus the double differential cross section of a mixture is described by

$$\sigma(E_0, E, \theta) = \frac{1}{4\pi}\sqrt{\frac{E}{E_0}}\,[2(1-x_D)\sigma_H(E_0,E,\theta) + 2x_D\sigma_D(E_0,E,\theta) + \sigma_O(E_0,E,\theta)] \qquad (3)$$

where E_0 and E are respectively the initial and final energy of the interaction, and $\sigma_A(E_0, E, \theta)$ is the free gas scattering function of atom A.

The input data necessary to perform Monte Carlo simulations require prior knowledge of detector and material parameters. In the following we will describe them, and how to obtain the desired information

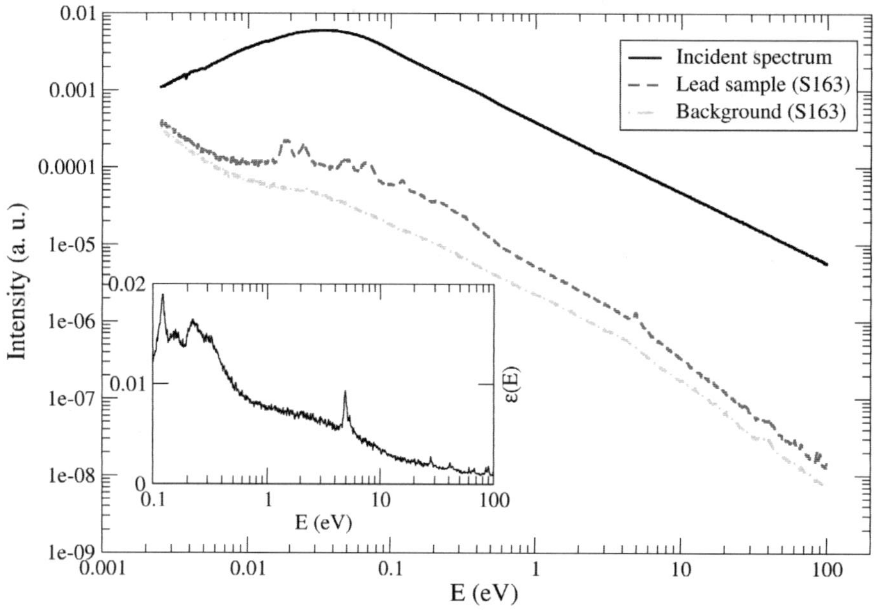

Figure 2. Incident neutron spectrum measured with detector S2 shown together with a lead sample spectrum measured with detector S163, located at 75.6 cm of the sample, at an angle of 42.8^o in the forward direction. Also, a background spectrum measured without sample. Inset: the efficiency of detector S163 determined from Eq. (2).

(a) Incident spectrum corrected by detector efficiency. We already described the measurement of the incident spectrum. To get $\Phi(E_0)$ employed in Eq. (1) the measured spectrum must be divided by the detector efficiency. The efficiency curve of the ^{6}Li-doped glass detector S2, based on calibrations previously done at VESUVIO is $\varepsilon_{S2}(E) = 1 - \exp\left[-\frac{0.051}{\sqrt{E}}\right]$ where the energy E must be expressed in eV.

(b) Efficiency of the analyzed detector. We determined it as described in §4.1.2. In the inset of Fig. 2 we show the efficiency of detector S163.

(c) Sample and filter geometry and materials. Sample inner and outer diameters and lids thickness are required parameters. Also the tabulated total cross sections as a function of energy of the sample and the container must be given as input.

(d) Instrument parameters. These parameters include the incident beam energy and its transverse section dimensions. The gold filter analyzer thicknesses are also required.

Journal of Physics: Conference Series **571** (2014) 012008 doi:10.1088/1742-6596/571/1/012008

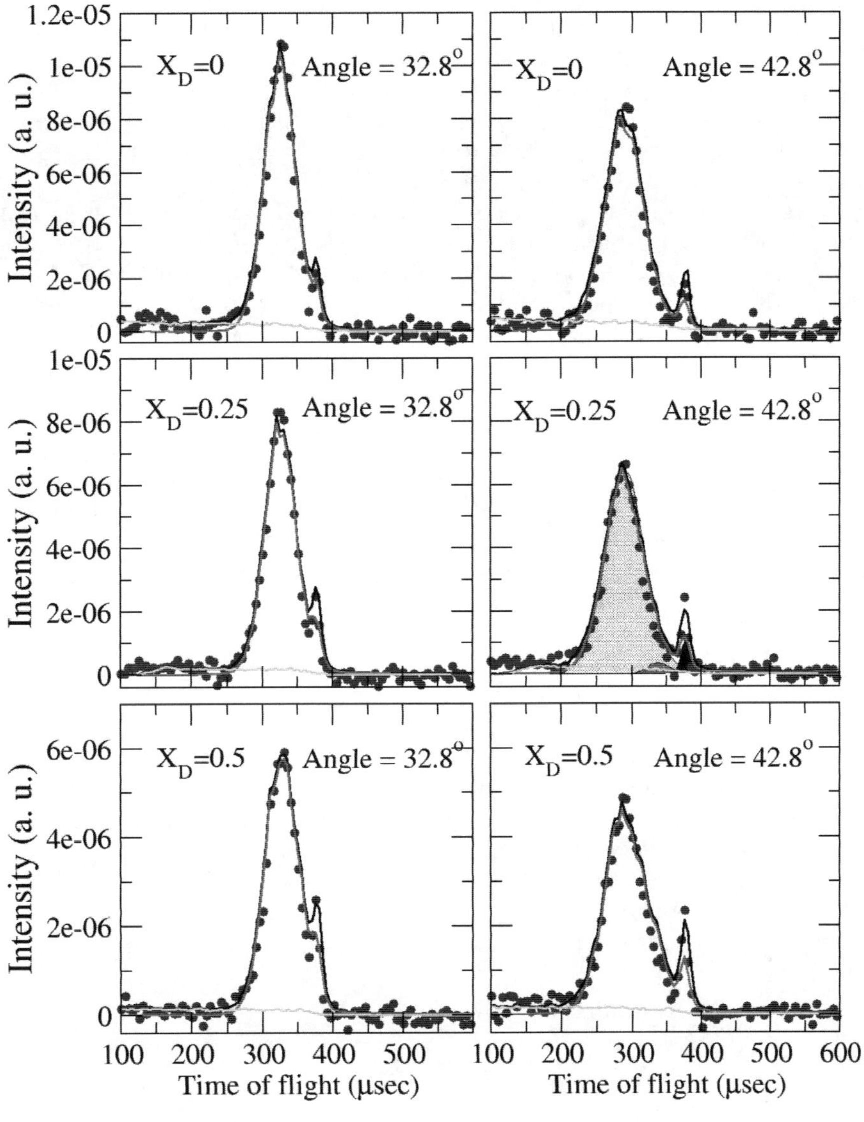

Figure 3. Experimental results for H_2O and the water mixtures H_2O/D_2O for $x_D = 0.25$ and 0.5 (blue circles) at two different angles in forward scattering, compared with the Monte Carlo simulation results. Red line: single scattering, green line: multiple scattering, black line: total scattering. Also shown contributions of H (light shaded area), D (medium shaded area), and O (dark shaded area), for $x_D = 0.25$ at 42.8^o

(e) Filter analyzer absorption cross section as a function of energy. Absorption cross section tables can be obtained from Ref. [15]. As analyzed in Ref. [16] a good description of the time-of-flight profiles requires a complete table of the filter analyzer absorption.

(f) Total cross section of the sample materials as a function of energy. These data must be tabulated in an energy range wide enough to cover not only the energies corresponding to the incident spectrum (epithermal energies), but also to consider the energy transfers after a number multiple scattering steps (typically thermal energies). As shown in the next section these data can be provided by the information generated by the transmission monitor.

Fig. 3 shows the Monte Carlo results compared with the experimental data for the two selected angles. With red line we show singly scattered neutron contribution, with green line the multiple scattering and with black line the total scattering.

The Monte Carlo results are used to define the attenuation and multiple scattering factors that must be used to correct the experimental data. The attenuation factor $H(t)$ is defined as ratio of detected singly-scattered neutrons to the total singly-scattered neutrons, and the multiple scattering factor $M(t)$ is defined as the ratio of detected singly-scattered neutrons to the total scattered neutrons[12]. The experimental data are then corrected by

$$E_{corr}(t) = \frac{M(t)E(t)}{H(t)} \tag{4}$$

Fig. 4 shows the uncorrected and corrected data that result after applying Eq. (4). In the inset we show the attenuation factors, and the multiple scattering factors. The variation in peak intensities after corrections will influence the final quantitative results. A detailed study of the observed intensities will be presented in a future publication.

4.3. Transmission monitor

We explored the possibility to measure total cross sections with VESUVIO transmission monitor S2. To this end we measured the transmitted beam of each sample ($S(E)$), and the empty cell ($F(E)$). As in the former measurements the spectra were normalized by monitor counts of detector S1. To assess the magnitude of the background counts (i.e. the number of neutrons counted by S2 that do not come from the incident beam), we measured the transmitted beam of a Cadmium sheet 1mm thick. In the thermal region, the magnitude of this spectrum was between 2 and 3 orders of magnitude lower than the open beam, so we neglected its contribution.

The transmission is determined from

$$T(E) = \frac{S(E)}{F(E)} \tag{5}$$

and the total cross section $\sigma_{tot}(E)$ results from [17]

$$T(E) = \exp(-nT\sigma_{tot}(E)) \tag{6}$$

where is n the number density of the sample and T its thickness. The same samples prepared at ISIS were taken to the Bariloche LINAC laboratory, where control measurements were carried out, following the same procedure described in Ref. [17].

The results are shown in Fig. 5, where the total cross sections of the three mixtures measured in both laboratories are compared. For illustrative purposes, in the inset we show the 'open beam' (empty cell) and the 'sample transmitted' spectra measured at VESUVIO, for the H_2O sample. The agreement is remarkable, and the comparison serves to validate the measurements at VESUVIO since the Bariloche facility is customarily employed in this kind of experiments and its setup optimized for this purpose.

Journal of Physics: Conference Series **571** (2014) 012008 doi:10.1088/1742-6596/571/1/012008

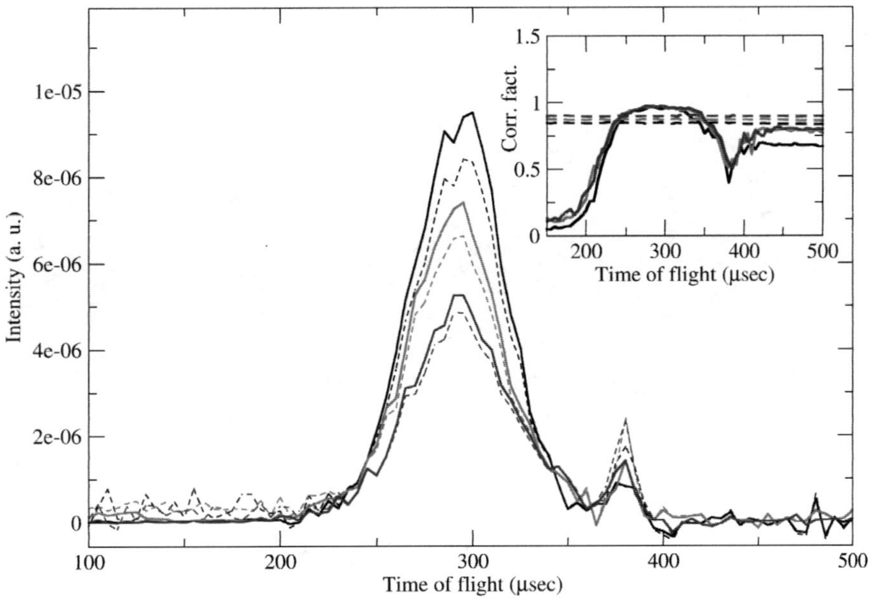

Figure 4. Uncorrected (dashed line) and corrected data (solid line) for H_2O (black lines) and the water mixtures H_2O/D_2O for $x_D = 0.25$ (red lines) and 0.5 (blue lines). Inset, the attenuation factors (dashed lines) and the multiple scattering factors (solid lines), with the same color convention.

5. Conclusions

In this paper we have shown two different aspects in the use and the analysis of the experimental information provided by VESUVIO. First, the use of information provided by single detectors in the time-of-flight scale is useful for quantitative analysis of atoms present in the sample. The steps that comprise analysis shown include multiple scattering attenuation and detector efficiency corrections, lead to a quantitative analysis in the way shown in Ref. [18]. In the second example, we showed that the transmission monitor S2 can be used to determine total cross sections of good quality, an application that until today had not been reported.

Future work under current development, shows the possibility to determine in a single experiment the composition of a sample and the total cross section of an element. The possibility of developing this type of experiments, will be of interest for those fields requiring the use of updated total cross sections, particularly Nuclear Engineering, that often use outdated information due to the lack of experimental facilities that provide it. In this sense, we envisage the possibility of future demand of information at different conditions of pressure and temperature.

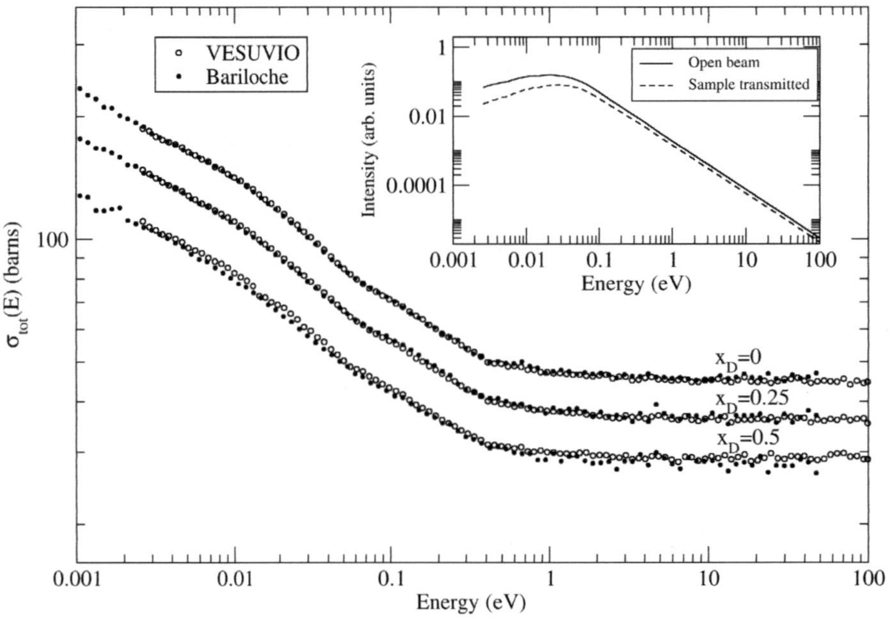

Figure 5. Total cross sections of the three water mixtures measured at VESUVIO (open symbols) compared with measurements made on the same mixtures made at the Bariloche LINAC. Inset: Open beam and transmitted sample spectra at detector S2 of VESUVIO.

Acknowledgments
We gratefully acknowledge L. Capararo and F. Moreira for the technical support, and M. Schneebeli, P. DAvanzo and A. Mancilla for the LINAC operation at Bariloche, and to Rutherford Appleton Laboratory (J. Tomkinson) for the invitation to share the beam time to make the experiments. Work partially supported by ANPCyT (PICT 2011 - 534) and CONICET (PIP 11220110100552) (Argentina).

References
[1] A. Pietropaolo, R. Senesi, Physics Reports 508 (2011) 4590.
[2] M. Krzystyniak, S. E. Richards, A. G. Seel and F. Fernández-Alonso Phys. Rev. B 88, (2013), 184304.
[3] J. J. Blostein, J. Dawidowski, and J. R. Granada, Phys. Rev. B 71, (2005), 054105
[4] J. Mayers and G. Reiter, Meas. Sci. Technol. 23 (2012) 045902.
[5] J. R. Granada, J. R. Santisteban, J. Dawidowski and R. E. Mayer, Physics Procedia 26 (2012) 108
[6] C. Andreani, A. Pietropaolo, R. Senesi, G. Gorini, E. Perelli-Cippo, M. Tardocchi, N. Rhodes, and E. M. Schooneveld, Applied Physics Letters 85, (2004), 5454.
[7] J. G. Powles, Mol. Phys. 26, 1352, (1976).
[8] L.A. Rodríguez Palomino, J.J. Blostein and J. Dawidowski, Journal of Instrumentation 8, (2013), P08016.
[9] F. A. Akeroyd, R. L. Ashworth, S. D. Johnston, J. M. Martin, C. M. Moreton-Smith, D. S. Sivia, Rutherford - Appleton : Laboratory Technical Report RAL-TR-2000-002 (2000).

[10] Calculation of background effects on the VESUVIO eV spectrometer, J Mayers Rutherford - Appleton : Laboratory Technical Report RAL-TR-2010-012.
[11] R. Bedogni, A. Esposito, C. Andreani, R. Senesi, M. P. De Pascale, P. Picozza, A. Pietropaolo, G. Gorini, C. D. Frost, S. Ansell, Nucl. Instr. and Meth. A 612 (2009) 143.
[12] J. Dawidowski, G. J. Cuello, L. A. Rodríguez Palomino, Neutron Diffraction, ISBN 979-953-307-475-6, Book edited by: Dr Khidirov, InTech - Open Access Publisher (2012).
[13] J.J. Blostein, J. Dawidowski, J.R. Granada , Physica B 385386 (2006) 69 .
[14] J.R. Granada, Phys. Rev. B 31 (1985) 4167.
[15] ENDF/B-VI, IAEA Nuclear Data Section, 1997.
[16] J. J. Blostein, J. Dawidowski, J.R. Granada, Physica B 304, (2001), 357.
[17] J. J. Blostein, J. Dawidowski, S. A. Ibáñez, and J. R. Granada, PhysRevLett. 90, (2003), 105302 .
[18] J. J. Blostein, L. A. Rodríguez Palomino, and J. Dawidowski Phys. Rev. Lett. 102, (2009) 097401.

Journal of Physics: Conference Series **571** (2014) 012009 doi:10.1088/1742-6596/571/1/012009

VESUVIO Data Analysis Goes MANTID

S. Jackson[1,2], M. Krzystyniak[1,3], A.G. Seel [1], M. Gigg[1,4], S.E. Richards[1,3], and F. Fernandez-Alonso[1,5]

[1] ISIS Facility, Rutherford Appleton Laboratory, Chilton, Didcot, Oxfordshire OX11 0QX, United Kingdom
[2] Department of Computer Science, University of Aberystwyth, Penglais, Aberystwyth, SY23 3DB, United Kingdom
[3] School of Science and Technology, Nottingham Trent University, Clifton Campus, Nottingham NG11 8NS, United Kingdom
[4] Tessella Ltd., Abingdon, Oxfordshire, UK
[5] Department of Physics and Astronomy, University College London, Gower Street, London WC1E 6BT, United Kingdom

Abstract. This paper describes ongoing efforts to implement the reduction and analysis of neutron Compton scattering data within the MANTID framework. Recently, extensive work has been carried out to integrate the bespoke data reduction and analysis routines written for VESUVIO with the MANTID framework. While the programs described in this document are designed to replicate the functionality of the Fortran and Genie routines already in use, most of them have been written from scratch and are not based on the original code base.

1. Introduction

Neutron Compton scattering (NCS) may be regarded as a mass spectroscopic technique in which each atomic mass contributes to the overall time-of-flight (TOF) spectrum in the form of a Doppler-broadened recoil peak. The width of such recoil peak in the energy domain can be related to the momentum distribution of the target nucleus prior to the scattering effect. NCS has been successfully applied in a wide variety of condensed matter systems and molecules. This includes fluid and solid hydrogen [1, 2, 3]; water [4, 5], proton glasses [6], biological molecules [7, 8], hydrogen storage materials [9, 10], helium (^{3}He and ^{4}He) [11, 12, 13, 14, 15, 16], noble gases in solid and fluid phases [17, 18], graphite [19], metal hydrides [20], etc.

NCS data-reduction schemes can be conveniently subdivided into two groups: (i) those that aim solely at the determination of scattered intensities [21, 22], and (ii) those that serve the purpose of the determination of kinetic energies and momentum distributions of individual nuclei in a mass-selective manner [23, 24, 25, 26, 27, 28]. The use of existing NCS data treatment schemes usually requires remote access to a variety of different computing environments with routines compiled for different operational systems and hardware infrastructures. These requirements substantially increase the complexity and time for the data analysis and reduction stage. The first step towards decreasing this complexity was made by extending the capabilities of the first generation of NCS data reduction routines by translating them into a set of MATLAB routines within a common NCS toolbox where computation, visualization, and programming can be completed within the same environment. Nevertheless, both FORTRAN and MATLAB versions of the NCS software have been developed and maintained in a closed environment

Journal of Physics: Conference Series **571** (2014) 012009 doi:10.1088/1742-6596/571/1/012009

resulting with suboptimal code development. Therefore, consolidation and translation of the NCS data reduction routines into the MANTID framework has become a prerequisite both for further code and technique development. MANTID, with extensive use of PYTHON and C++, uses a common interface which is seamless between the facility, beam line and home institution when performing the final analysis and fitting of the data [29]. Furthermore, this approach increases transparency between the developer and user community. This has been the main motivation behind the work presented here. In this paper, we briefly describe VESUVIO data-reduction and analysis as implemented in Fortran, MATLAB, and MANTID, and use the example of zirconium hydride as a test bed to demonstrate their equivalence. Examples of the NCS data-reduction routines using the MANTID graphical user interface and PYTHON scripts are presented. In the final section additional functionalities are discussed in the context of their potential of providing new insights in the analysis of NCS data.

2. Neutron Compton scattering and the impulse approximation

In NCS, the energy and momentum transfers imparted by the neutron to the target nucleus are so high that the scattering process can be treated within the (IA) [30, 31, 23]. The IA effectively treats NCS as billiard-ball scattering with conservation of kinetic energy and momentum of the neutron and a given (struck) nucleus. Under the IA, the scattering function, $S_M(\mathbf{q}, \omega)$ for a given nucleus of mass M reduces to a single peak centered at the recoil energy $E_r = \hbar^2 Q^2/2M$. In this limit, $S_M(\mathbf{q}, \omega)$ has the form [23]:

$$S_M(\mathbf{q}, \omega) = \int n_M(\mathbf{p}) \delta\left(\omega + \frac{p^2}{2M} - \frac{(\mathbf{p}+\mathbf{q})^2}{2M}\right) \tag{1}$$

where $n_M(\mathbf{p})$ is the nuclear radial momentum distribution for mass M.
Consequently we can write $S_M(\mathbf{q}, \omega) = \frac{M}{\hbar^2 q} J(y, \hat{q})$, where:

$$y = \mathbf{p} \cdot \hat{\mathbf{q}} = \frac{M}{\hbar^2 q}(\omega - \omega_r) = \frac{M}{\hbar^2 q}\left(\omega - \frac{\hbar^2 q^2}{2M}\right) \tag{2}$$

where $\hat{\mathbf{q}}$ is a unit vector in the direction of the momentum transfer. Hereinafter, momentum transfer will be given in units of $\mathring{A}^{-1}$, energy transfer ω in meV, and atomic mass in amu. With this choice of units, Plank's constant is given by $\hbar = 2.0446$ (meV amu)$^{1/2}$ Å.

$J(y, \hat{\mathbf{q}})$ is the so called neutron Compton profile [31, 30] and is formally the result of a Radon transformation of the radial momentum distribution $n_M(\mathbf{p})$. The relationship given by Eq. (2) is formally known as 'y-scaling' or the 'West-Scaling' [32, 31, 23]. It implies that q and ω are no longer independent variables and that for samples with no preferred orientation any scan in $\mathbf{q} - \omega$ space crossing the recoil line, defined by the relation $\omega_r = \frac{\hbar^2 q^2}{2M}$, gives a measurement of the neutron Compton profile $J(y, \hat{\mathbf{q}})$. For a harmonically bound isotropic system where the momentum distribution $n_M(\mathbf{p})$ depends only on the magnitude of the radial momentum, $p = |\mathbf{p}|$, and the neutron Compton profile $J(y, \hat{\mathbf{q}})$ in the IA limit, hereinafter denoted as $J_{IA}(y)$, can be written as a normalized Gaussian [33, 34, 35, 36, 23] of the form

$$J_{IA}(y) = \frac{1}{\sqrt{2\pi\sigma_p^2}} \exp\left(\frac{-y^2}{2\sigma_p^2}\right) \tag{3}$$

where σ_p is the standard deviation. Equation is exactly correct when the atoms are bound by isotropic harmonic forces. In real data, the anisotropy and anharmonicity of the local binding potential both give a non-Gaussian $J_{IA}(y)$. Equation (3) is generally good when small atomic displacement from the mean position sample the harmonic region. The current resolution

Journal of Physics: Conference Series **571** (2014) 012009 doi:10.1088/1742-6596/571/1/012009

and count rates on VESUVIO make the measurement of any small deviations from Gaussian behaviour for $M >\sim 20$ impractical.

Even when the IA is not satisfied, $y$scaling can still be an appropriate starting point. For finite values of q, corrections to the IA known as 'final-state effects' (FSEs) are usually expressed using y scaling [31, 36, 23]. To account for FSEs the method of Sears is routinely incorporated into standard NCS data treatments [31, 36, 23] by expressing the measured neutron Compton profile $J(y)$ to first order as

$$J(y) = J_{IA}(y) + J_{FSE}(y) = J_{IA}(y) - \frac{M\langle\nabla^2 V\rangle}{36\hbar^2 q}\frac{d^3}{dy^3}J_{IA}(y) + ... \tag{4}$$

where $J_{IA}(y)$ is the IA result. $\langle\nabla^2 V\rangle$ is the mean value of the Laplacian of the potential energy of the nucleus expressed in meVÅ^{-2} (cf. Refs.[19, 36, 23]. In an NCS experiment performed on an isotropic sample within the IA limit, a given NCS profile $J_{IA}(y)$ can be related to a spherically averaged three-dimensional nuclear momentum distribution (NMD) $n_{IA}(p)$ by the following expression [23]

$$n_{IA}(p) = \frac{-1}{2\pi y}\frac{\partial J_{IA}(y)}{\partial y}\bigg|_{y=p} \tag{5}$$

where the radial NMD $n_{IA}(p)$ can be represented as

$$n_{IA}(p) = \frac{\exp\left(\frac{-p^2}{2\sigma^2}\right)}{\left(\sigma\sqrt{2\pi}\right)^3}\sum_n c_n(-1)^n L_n^{1/2}\left(\frac{p^2}{2\sigma^2}\right) \tag{6}$$

where $L_n^{1/2}$ denotes a generalised Laguerre polynomial. Equation (6) can be used for the reconstruction of $n_{IA}(p)$ from the measured $J(y)$ after subtraction of the FSE contribution $J_{IA}(y) = J(y) - J_{FSE}$ where

$$J_{IA}(y) = \frac{\exp\left(\frac{-y^2}{2\sigma^2}\right)}{\left(\sigma\sqrt{2\pi}\right)}\sum_n \frac{c_n}{2^{2n}n!}H_{2n}\left(\frac{y}{\sigma\sqrt{2}}\right) \tag{7}$$

where H_{2n} denotes a Hermite polynomial. In the above expression, the coefficients c_n are identical to the expansion coefficients for the radial momentum distribution, which is tantamount to an inversion procedure between $J_{IA}(y)$ and $n_{IA}(p)$. The expansions are in fact over complete: σ can be arbitrary and the series in Eq. (7) will still converge satisfactorily to an arbitrary function [27]. However, if the first term is omitted in the fitting (that is c_1 is fixed at zero), σ necessarily determines the kinetic energy, independently of any other terms in the series.

3. Count rate on an inverse-geometry instrument

On an inverse geometry instrument like VESUVIO at ISIS [27], the time-of-flight (TOF), t, of each detected neutron is determined by

$$t = \frac{L_0}{v_0} + \frac{L_1}{v_1} \tag{8}$$

where L_0 is the source-sample distance, L_1 is the sample-detector distance, $v_0 = \sqrt{2E_0}/m$ is the incident neutron velocity, and E_0 the incident neutron energy, $v_1 = \sqrt{2E_1}/m$, the velocity corresponding to detected (and thus fixed) neutron energy E_1, and m is the neutron mass. The incident energy E_0, can be calculated from the measured TOF, t (Eq. 8), given L_0, L_1, and E_1. Hence the energy transfer

$$\omega = E_0 - E_1 \tag{9}$$

and the momentum transfer

$$q = \sqrt{2m}\left(E_0 + E_1 - 2cos\theta\sqrt{E_0E_1}\right)^{1/2} \tag{10}$$

can both be determined from t.

The count rate as a function of t on an inverse-geometry neutron spectrometer can be written as [27]

$$C\left(E_0\right)\mathrm{d}E_0 = 2\left(\frac{2}{m}\right)^{\frac{1}{2}}\frac{\left[E_0\right]^{3/2}}{L_0}I\left(E_0\right)D\left(E_1\right)N\frac{\mathrm{d}^2\sigma}{\mathrm{d}E_1\mathrm{d}\Omega}\mathrm{d}\Omega \tag{11}$$

where $I\left(E_0\right)\mathrm{d}E_0$ is the number of incident neutrons per second per unit area with energies between E_0 and $E_1+\mathrm{d}E_0$. $D\left(E_1\right)$ is the probability that a neutron energy E_1 is detected, $\frac{\mathrm{d}^2\sigma}{\mathrm{d}E_1\mathrm{d}\Omega}$ is the double differential cross-section, N is the number of atoms in the beam, and $\mathrm{d}\Omega$ is the detector solid angle.

The spectra are, however, recorded and represented as a function of t instead of E_0 with $C(t) = C(E_0)dE_0/dt$. The Jacobian dE_0/dt is given by the following expression [36]:

$$\frac{dE_0}{dt} = \frac{2}{L_0}\sqrt{\frac{2}{m}}E_0^{3/2} \tag{12}$$

Using the Jacobian dE_0/dt the count rate $C_\theta\left(t\right)$ in a given TOF channel t becomes [36] :

$$C_\theta\left(t\right) = \left[\frac{E_0I\left(E_0\right)}{q}\right]\left(t\right)\times A'MI_MJ_M\left[y_M\left(t\right)\right] \tag{13}$$

where A' is another mass-independent experimental constant.

If atoms of different masses M are present in the sample, the total count rate can be generalised to [36] :

$$C_\theta\left(t\right) = A'\left[\frac{E_0I\left(E_0\right)}{q}\right]\left(t\right)\times\sum_{n=1}^{M}I_nM_nJ_n\left[y_n\left(t\right)\right] \tag{14}$$

where both the experimental constant A' and the factor $\left[\frac{E_0I(E_0)}{q}\right]\left(t\right)$ are mass independent and can be written in front of the sum over M. The integrated peak intensity I_M for a given mass M is given by $I_M = AN_M\sigma_M$, where $\sigma_M = 4\pi b_M^2$ is the total bound scattering cross section and N_M is the number density. [36]

4. Instrument resolution

In a real experimental situation the measured count rate $C_\theta(t)$ for every TOF, t, is an average over the possible values of all geometrical parameters characterising neutron trajectories and scattering angles as well as the final neutron energy weighted by its probability of occurrence. In case of NCS data reduction the most frequently adopted formalism to account for the effects of instrument resolution is to replace the multiple integrals over distributions of the above-mentioned parameters that would have been required to be calculated for each data point in a TOF spectrum for a given mass by convolution with a mass-dependent resolution function. In the framework of the so-called 'convolution approximation' (CA) it is assumed that a TOF spectrum for a given mass is described by a convolution of a mass-dependent resolution function of the instrument $R\left[y_M\left(t\right)\right]$ and the neutron Compton profile of the scatterer $J[y_M(t)]$ [36]. Typically, the resolution function $R\left[y_M\left(t\right)\right]$ is assumed to be described by a Voigt line shape as it is the case for the gold foil used on VESUVIO [36]. The convolution is performed in y-space for each mass M and then the sum over contributions from different masses is performed to account for the total count rate recorded in a given TOF spectrum. Thus, for a total of N different masses present in the sample, the total count-rate at a fixed scattering angle θ, $C_\theta(t)$ is given by [36] :

$$C_\theta(t) = A' \left[\frac{E_0 I(E_0)}{q}\right](t) \times \sum_{n=1}^{m} I_n M_n J_n\left[y_n(t)\right] \otimes R_n\left[y_n(t)\right] \quad (15)$$

where A' is a mass-independent experimental constant and the mass-independent factor $\left[\frac{E_0 I(E_0)}{q}\right](t)$ depends on the incident neutron spectrum, $I\left[E_0(t)\right]$ the initial neutron energy, $E_0(t)$, and the momentum transfer $q(t)$, all functions of the TOF, t [36]. The difference between the instrument resolution function implemented in [27] and the instrument resolution function implemented in the NCS reduction routine used in the MANTID environment is a convolution in y-space rather than t [27].

FSE contributions to both Gaussian and non-Gaussian J_M are accounted for by including terms proportional to $1/qH_3$. The resulting expression to fit the entire TOF spectrum is of the following form [37] :

$$C_\theta(t) = A' \left[\frac{E_0 I(E_0)}{q}\right](t) \times \left[J_H(x_H) \otimes R_H(x_H) + \sum_{M \neq H} J_M(x_M) \otimes R_M(x_M)\right] \quad (16)$$

where J_H is the longitudinal nuclear momentum distribution for the first mass (i. e. the mass for which the recoil peak is centered at the lowest TOF value)

$$J_H(x_H) = \frac{\exp\left(-x_H^2\right)}{\sqrt{2\pi_H^2}} \left(1 + \frac{c_4}{32} H_4(x_H) - \frac{k}{q} H_3(x_H)\right) \quad (17)$$

and J_M is the longitudinal nuclear momentum distribution for any other subsequent mass M

$$J_M(x_M) = \frac{\exp\left(-x_M^2\right)}{\sqrt{2\pi_M^2}} \left(1 - \frac{k}{q} H_3(x_H)\right) \quad (18)$$

with $x_M = \frac{y_M - y0_M}{\sqrt{2\sigma_M^2}}$, where $y0_M$ is the shift of the position of the maximum of a nuclear momentum distribution from the center of the recoil line and σ_M is the standard deviation of the momentum distribution for a mass M. The FSE expansion in the limit of a local effective harmonic potential for a given mass M leads to a very useful expression of the FSE coefficient k in the expansion given by $k = \sigma\frac{\sqrt{2}}{12}$ [37]. Thus, fitting of the recoil peaks in a harmonic approximation does not require any extra fitting parameter to describe FSEs.

5. VESUVIO data analysis in MANTID

The MANTID development team in combination with VESUVIO instrument scientists have created a data analysis package to help aid with the analysis of VESUVIO data. This is still a work in progress and under heavy development. Currently it is in the form of a separate PYTHON module called *ncs.py* which provides a collection of helper functions and classes for data analysis which builds upon the existing MANTID framework. This module is publicly available in the MANTID GITHUB scripts repository in the `development\inelastic` folder [38]. The simplest way of including the NCS PYTHON module in MANTID is to place the files *ncs.py* and *ncs plotting.py* in the MANTID `Install\scripts` folder. This will allow the NCS PYTHON module to be imported directly into the MANTID script window.

Journal of Physics: Conference Series **571** (2014) 012009 doi:10.1088/1742-6596/571/1/012009

5.1. Data loading and pre-treatment options

The original data analysis package [39] used two separate commands to process the raw time-of-flight data, one for the front scattering detectors and one for the back scattering. In the MANTID implementation this has been replaced with a single algorithm called *LoadVesuvio* which performs all of the processing for raw files from the instrument. This includes options for handling each of the different foil positions and difference techniques available with VESUVIO [40, 36] and handles the summing multiple runs. It also contains a flag for summing each spectrum in the desired range into a single spectrum. The results of this loading operation are output as a single workspace in MANTID with units in time-of-flight which can be plotted using both MANTID built in visualisation tools and custom plotting commands described in Ref. [39](see Fig. 1).

Like all inelastic instruments at ISIS, VESUVIO uses time of flight measurements to determine the momentum and energy transfer to the neutron in each scattering event. The neutron counting chain assigns a time of flight to the arrival of each neutron at the detector according to Eq. (7). On VESUVIO the values of geometrical parameters defining the TOF, $L0$, $L1$, $v1$, $t0$, and the scattering angle, θ, can be characterised by mean values, with associated distributions about the mean. Given these mean values and the value of TOF assigned to the neutron, the incident velocity, $v0$, of each neutron detected and hence the momentum and energy transfer in the scattering process can be determined from Eq. (7). The distributions of these parameters about their mean values determine the resolution function of the instrument. The main task instrument calibration procedure is to determine these distributions and store them in the instrument parameter (IP) file.

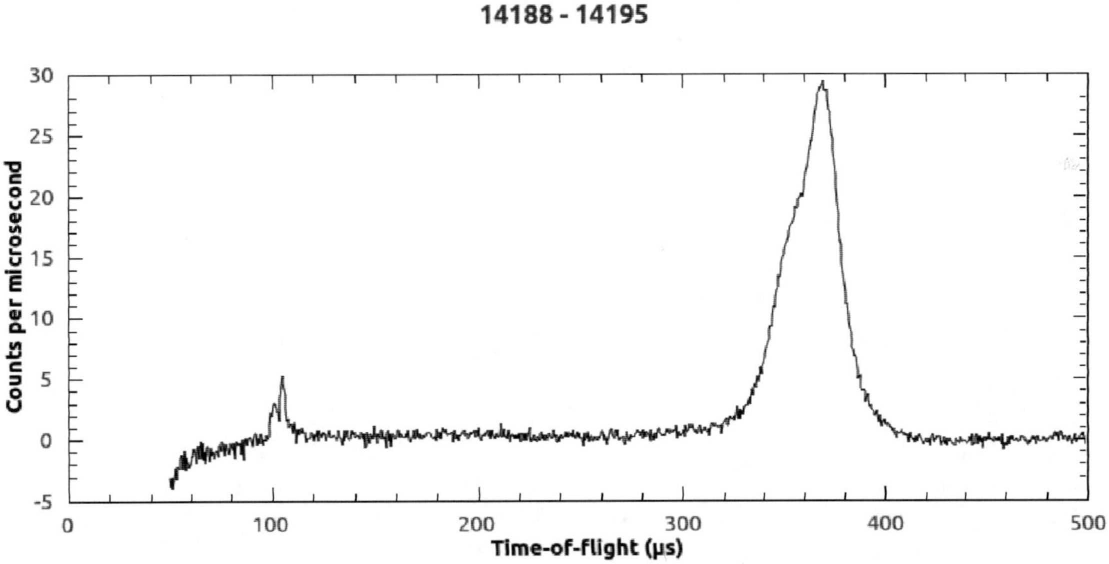

Figure 1. Plot of time-of-flight data from the sum runs 14188-14195 and the sum of detectors 40-134.

The parameter file is usually generated by the instrument scientist using a set of reference runs using the calibration program outlined in section 5.8 and does not need to be regenerated for every analysis session.

The data pre-treatment procedure in the MANTID NCS reduction scheme includes: definition of differencing technique for final-energy analysis (i.e., single or double differencing) [40, 41]

Journal of Physics: Conference Series **571** (2014) 012009 doi:10.1088/1742-6596/571/1/012009

followed by smoothing and truncation of data points. Residual γ-background correction, multiple-scattering correction and baseline fitting complete the data pre-treatment procedure.

5.2. Definition of differencing technique

$\gamma-$ray signals are detected on VESUVIO neutron detectors at forward-scattering angles [42]. These are made to function both as neutron detectors and energy selectors. In forward scattering, a foil-cycling (FC) technique is employed whereby two foils of the same resonance material are used, one stationary analyzer foil and one cycling foil that alternates in and out of the scattered neutron beam [40]. The energy E_1 of detected neutrons is by a large neutron absorption resonance peak at E_1= 4.9 eV. By placing a primary gold foil on the YAP detector surface, the foil absorbs neutrons of energy $E_1\approx$ 4.9 eV and emits a γ-ray cascade which can be detected, thereby determining both the TOF and the final energy of the neutron. These measurements are called 'foil-out' measurements because neutrons do not have to pass through the secondary gold foil. The γ-ray signal produced from the 'foil-out' measurements sits on a background, which may be due to γ-ray rays produced when neutrons initially scattered by the sample are then absorbed by the surrounding shielding. By positioning the secondary foil between the sample and the detector, the signal produced from these scattered neutrons are significantly reduced as neutrons of energy $E_1\approx 4.9eV$ have to first pass through the secondary foil (or filter) to reach the primary foil. Through the use of this secondary gold foil, the energy resolution is also improved and the background almost eliminated. The VESUVIO 'raw' data is the difference between these two counts ('foil-out' minus 'foil-in') [42, 43].

In backscattering, gold foils are placed in front of ^{6}Li-doped neutron detectors. In this case, a double-difference (DD) technique [41] is used to remove the background and sharpen the energy resolution. This method is based on a linear combination of two measurements recorded with filter foils of the same resonance material of different thickness. Both the FC and DD methods give a value of 4.897eV for the mean value of E_1 [44]. Detailed descriptions of the energy resolution functions can be found in Refs [41, 40, 45, 46].

On VESUVIO, typically the first step after loading the raw data in time-of-flight is to crop the workspace to a sensible range for data analysis. This can be done by using the MANTID *CropWorkspace* algorithm. The typical time-of-flight range is commensurate with the size of a data vector used in an FFT-based convolution routine used fitting, i.e., 50.0 - 562.0 μs (see Table 1).

Table 1. Example script showing how to load data and crop VESUVIO data using the MANTID PYTHON API.

```
1 import ncs
2 runs = "14188-14195"
3 spectra = "134"
4 diff_type="SingleDifference" # Allowed values=Single, Double, Thick
5 ip_file = "IP0004_10.par"
6 raw_ws = LoadVesuvio(Filename=runs, SpectrumList = spectra,
7 Mode=diff_type, InstrumentParFile = ip_file)
8 raw_ws = CropWorkspace(raw_ws, XMin=50.0, XMax=562.0)
```

The *ncs.py* module also provides a preprocessing function for further preparing the data for analysis. This method provides options for masking data points in the raw workspace given an error threshold as well as an option to smooth the workspace using the *SmoothData* algorithm. This requires a *FitOptions* object to be supplied (see section 5.5).

5.3. γ-background corrections in forward scattering

The γ-background correction utilizing the differencing techniques described above significantly reduces the γ-ray signals arising from the absorption of nuclei from the boron shielding on VESUVIO. However, the data from the YAP detectors in forward scattering still needs to be corrected for potential, sample-dependent (residual) γ-backgrounds because the intensity of the γ-background from the secondary foils in the instrument is dependent upon their position, which differs in the foil-in and foil-out measurements. This may result in spurious broad peaks in the baseline of the spectrum. Therefore the residual γ-background is calculated from the measured data [42] as:

$$\eta C_{out} - C_{in} = CA^2 + (B_{out} - B_{in}) \tag{19}$$

where C_{out} is the 'foil-out' count-rate from the primary gold foil, C_{in} is the 'foil-in' count-rate from the secondary gold foil, $A(E_1)$ is the analyzer-foil neutron absorption coefficient at final energy E_1, and η is the probability that the subsequent γ-ray cascade is registered by the YAP detector. B_{out} is the γ-background with the secondary foil in the 'out' position, and B_{in} is the background with the secondary foil in the 'in' position.

Corrections for the gamma background are implemented using the MANTID algorithm framework as an algorithm called *CalculateGammaBackground.* This takes a single workspace in time-of-flight, and a fit function describing the mass spectrum of the input data and a list of workspace indices to include as part of the correction. The fit function for the mass spectrum is typically one which has been created as part of the fit routines described in section 5.5. This algorithm results in two workspaces: one that contains the calculated background and a copy of the input time-of-flight workspace with the background subtracted. This function loosely corresponds to the *bcorr* command in the old data analysis package.

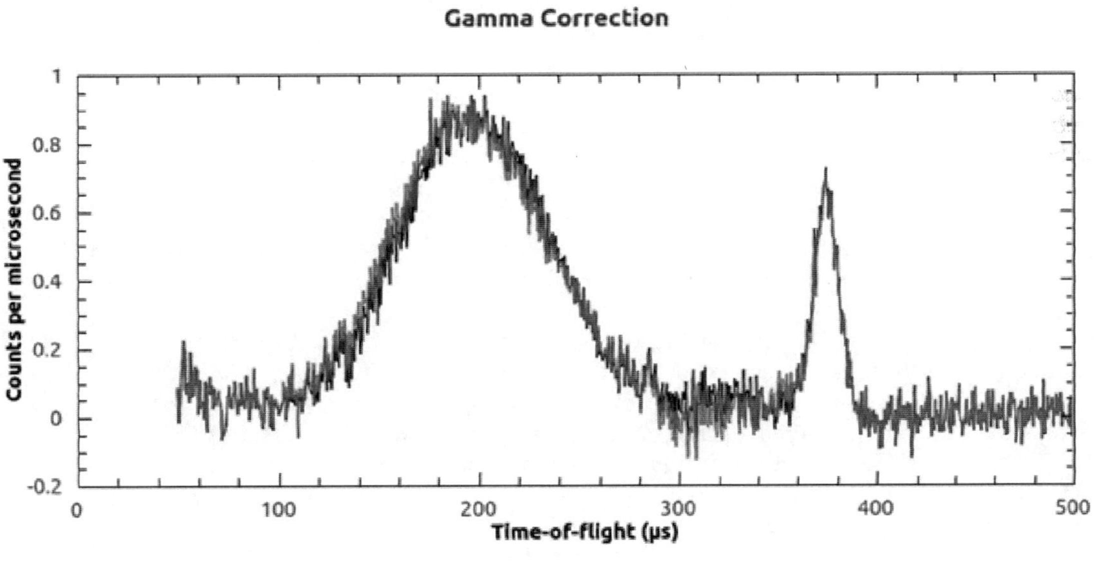

Figure 2. Plot of a Zirconium Hydride (ZrH_2) sample. Black is the uncorrected sample, red shows the same sample after gamma correction.

There are helper functions for this implemented in *ncs.py.* The gamma background can be computed from a workspace simply by using the *gamma correct* function and passing the raw workspace to correct, the required fitting options, and the parameter workspace produced from a fit (see section 5.5). Table 2 shows the code to perform a gamma correction on a workspace that has already been fitted. The result of the code application is shown in Figure 2.

Table 2. Example script for performing a gamma correction to the ZrH_2 sample using *ncs.py*.

```
1 ncs.preprocess(raw_ws, fit_options)
2 reduced_chi_square, params_ws = ncs.run_fit(raw_ws, fit_options)
3 ncs.display_fit_output(reduced_chi_square, params_ws, fit_options)
4 background, corrected = ncs.gamma_correct(raw_ws, fit_options, params_ws)
```

5.4. Multiple-scattering correction

Multiple scattering (MS) occurs because there is always a finite probability that neutrons will scatter more than once in the sample before being detected. MS must be subtracted from the TOF spectra before accurate results for the momentum distribution can be obtained. MS corrections are calculated using the Monte-Carlo method described in reference [47]. Currently, the multiple scattering corrections for VESUVIO in MANTID are under development and will be available in future releases.

5.5. Fitting Routine

The majority of the work to integrate VESUVIO into MANTID has been concerned with the development of the fitting procedures required to measure the neutron Compton profile following the theory described in Section 3. Development in this area has focussed on two major additions to the MANTID framework. First was the creation a new suite of fit functions which could accurately describe the results of a neutron Compton scattering experiment. The second was the creation of a collection of supporting data analysis functions which allow the user to easily set up a fit given the appropriate parameters for the sample and are available in the *ncs.py* module. These will eventually be folded into a GUI for VESUVIO (see section 5.9).

There are two major fit functions used in the analysis of VESUVIO data. The *GaussianComptonProfile* defines a function for fitting the simpler Gaussian approximation to mass peaks. The *GramChalierComptonProfile* is for the more complex fitting case described by Gram-Charlier expansion [5, 6]. Both functions are implemented as standard MANTID fit functions and are therefore directly accessible through the fit wizard. Currently, in order to run a fit a *FitOptions* object must be created from *ncs.py* and set up with the appropriate options for the sample being fitted. This includes a set of parameters for each mass describing it's atomic weight, the function it should be fitted with and any additional options specific to the model used. Table 3 shows an example of how to set up the parameters of a fit using the *ncs.py* *FitOptions* class. Table 4 lists all of the current fitting options available in the *FitOptions* class, while Table 5 describes each of the options for defining the masses to be fitted.

The simplest mode of analysis is to assume that each of the peaks in momentum space is of Gaussian shape [36]. Using the amplitudes and widths obtained for the fit the composition of the sample and the atomic kinetic energies may be determined. Table 6 shows a basic example of how to use the *FitOptions* class to set up fit for the aluminium and zirconium masses in the sample.

Fitting the data shown in Figure 2 with the parameters defined in Table 3 produces three workspaces: one containing the fit of the composite function and individual mass profiles to the raw data, a table workspace of parameters for the fit, and a table workspace for the normalised covariance matrix of the fit. These workspaces are in fact just the standard output from the

Journal of Physics: Conference Series **571** (2014) 012009 doi:10.1088/1742-6596/571/1/012009

Table 3. Example script for setting up a fit to the ZrH2 sample using *ncs.py*. The mass of aluminium and zirconium are fitted using the Gaussian fit function. In this example, the widths of each of the masses has been fixed using 'widths' attribute.

```
fit_options = ncs.FitOptions()
fit_options.workspace_index = 0
fit_options.bad_data_error = 1e6
mass1 = {'value':27.0, 'widths':[12,14.4, 16], 'function':'Gaussian'}
mass2 = {'value':91, 'widths':[24, 26.6, 28], 'function':'Gaussian'}
fit_options.masses = [mass1, mass2]
ncs.preprocess(raw_ws, fit_options)
reduced_chi_square, params_ws = ncs.run_fit(raw_ws, fit_options)
ncs.display_fit_output(reduced_chi_square, params_ws,fit_options)
```

Table 4. Table listing the different fitting options available in the *FitOptions* class.

Name	Description
smooth points	Number of data points to use in the *SmoothData* algorithm in the pre-processing function. If not set the data will not be smoothed.
bad data error	If set, the data will be masked if it falls outside of this error tolerance.
background function	The type of background to use in fitting. By default this is set to use the Polynomial background function.
background order	The order of the background to use.
masses	A dictionary of parameters for each mass supplied to the fit. This should include the atomic weight, function to fit (either *Gaussian* or *GramCharlier*) any other parameters required by the function.
constraints	Constraints on the intensity of each mass peak.
workspace index	The index in the workspace to fit to.
output prefix	String which is prefixed to output workspaces created from the fitting.
global fit	Whether or not to perform a multi-dataset (global) fitting of the data.

MantidFit algorithm. A plot of the fit using the Gaussian approximation function to two mass peaks for aluminium and zirconium is shown in Figure 3. Comparing this to the original example in Ref. [44] shows a good correlation between implementations.

The fitted amplitudes and widths of the sample can be examined by looking at the parameters table produced by the fit. In the original VMS data analysis package a routine called *ke* was used to calculate the kinetic energy in meV and K and the fundamental frequency in meV and cm^{-1} [39]. There is currently no direct implementation of this program in MANTID

The momentum distribution of atoms has a strictly Gaussian shape only when bound by isotropic harmonic forces. The shape of the nuclear momentum distribution is of interest as this contains information about the anisotropy of the binding and anharmonic effects [23]. In this case the data can be fitted using the Gram Charlier expansion [28]. Setting up such a fit is

Table 5. . Table listing the options for defining a mass to be fitted.

Name	**Description**
value	The atomic weight of the mass in amu
widths	The width of the peak. If this is set to a single number the width will be fixed to that value. If it is set to an iterable (tuple or list) of three values the middle value is the starting value and the first and last values are the constraints on the parameters fit.
function	The function to use to fit this mass. Options are currently either *Gaussian* or *GramCharlier*.
hermite coeffs	Only applicable when using *GramCharlier*. This is a list of coefficients for each mass being fitted. If set to one the term is included within the fit. If set to zero it is excluded from the fit.
k free	Only applicable when using *GramCharlier*. If this flag is set to true the FSE ("Final State Effects") coefficient is not tied to a value. If false it is set to a value dependent on the sears flag.
sears flag	Only applicable when using *GramCharlier*. If the k free flag is set to false, this flag will control what the FSE coefficient gets tied too. If set to one it is tied to $\sigma\frac{\sqrt{2}}{12}$ which is harmonic limit of the magnitude of the FSEs.

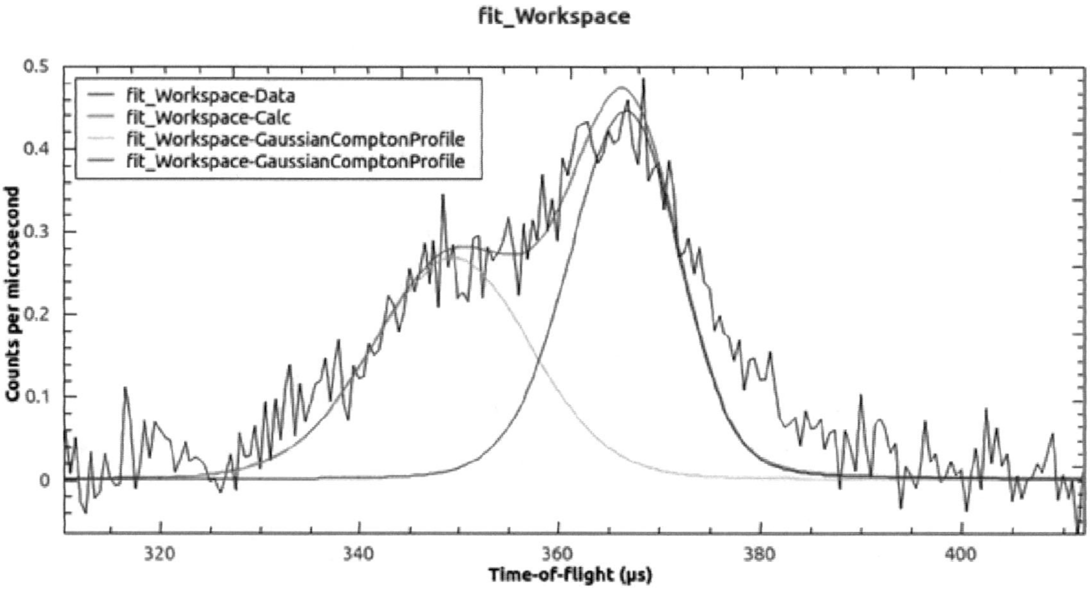

Figure 3. . Plot of the fitted time-of-flight data from the sum runs 14188-14195 at the scattering angle of 163 degrees for a sample of ZrH_2. The black line shows the original data, the red line shows the fit to the data is in red, and the individual contributing.

Journal of Physics: Conference Series **571** (2014) 012009 doi:10.1088/1742-6596/571/1/012009

very similar to the procedure described for the Gaussian approximation. Each mass being fitted using the expansion requires some additional parameters describing the Hermite coefficients to be used in the fit and what the Final State Effects (FSE) coefficient should be tied too. Table 6 shows an example of setting up such a fit using the data for the ZrH_2 sample.

Table 6. Example script for setting up a fit to the ZrH_2 sample using *ncs.py*. In this example, the widths of each of the masses except hydrogen has been fixed using 'widths' attribute.

```
fit_options = ncs.FitOptions()
fit_options.workspace_index = 0
fit_options.bad_data_error = 1e6
mass1 = {'value':1.0079, 'widths':[2,5,7], 'function':'GramCharlier',
'hermite_coeffs':[1,0,0],'k_free':False, 'sears_flag':1}
mass2 = {'value':27.0, 'widths':14.4, 'function':'Gaussian'}
mass3 = {'value':91, 'widths':26.6, 'function':'Gaussian'}
fit_options.masses = [mass1, mass2, mass3]
ncs.preprocess(raw_ws, fit_options)
reduced_chi_square, params_ws = ncs.run_fit(raw_ws, fit_options)
ncs.display_fit_output(reduced_chi_square, params_ws, fit_options)
```

In order to fit the data to determine the momentum distribution the raw data needs to first be normalised by the area of the peak. This removes the effects of different detector efficiencies and solid angles. The MANTID algorithm *NormaliseByPeakArea* has been created to handle this step. This algorithm effectively directly replaces the implementation of the *isofile* command in the original data analysis package [39]. This algorithm will:

(i) Transform the fitted data to y-space using the algorithm *ConvertToYSpace* for a single mass.

(ii) Fit the peak in y-space.

(iii) Divide the time of flight data by the area of the fitted peak in momentum space.

The *NormaliseByPeakArea* algorithm (see Table 7) produces four workspaces:

- *OutputWorkspace* - This workspace is the time-of-flight workspace supplied as input to the algorithm normalised by the area of the fitted peak.
- *YSpaceDataWorkspace* - This is the time-of-flight workspace converted to y-space using the *ConvertToYSpace* algorithm.
- *FittedWorkspace* - The values of the fit to the data.
- *SymmetrisedWorkspace* - The time-of-flight workspace converted to y-space and symmetrised about zero. As the momentum distribution should be symmetric about the origin, this workspace should match the *YSpaceDataWorkspace*.

Table 7. Example PYTHON code showing how to run the NormaliseByPeakArea algorithm.

```
normalised, yspace, fitted, symmetrised =
NormaliseByPeakArea(InputWorkspace=raw_ws, Mass=1.0079,Sum=True)
```

The final step in analysis is to simultaneously fit data across all detectors using the normalised workspace created as part of the previous section while keeping the widths of each peak tied (see Figure 4). This roughly emulates the program *isofileu* described in Ref. [39]. This global fitting option is still currently under development.

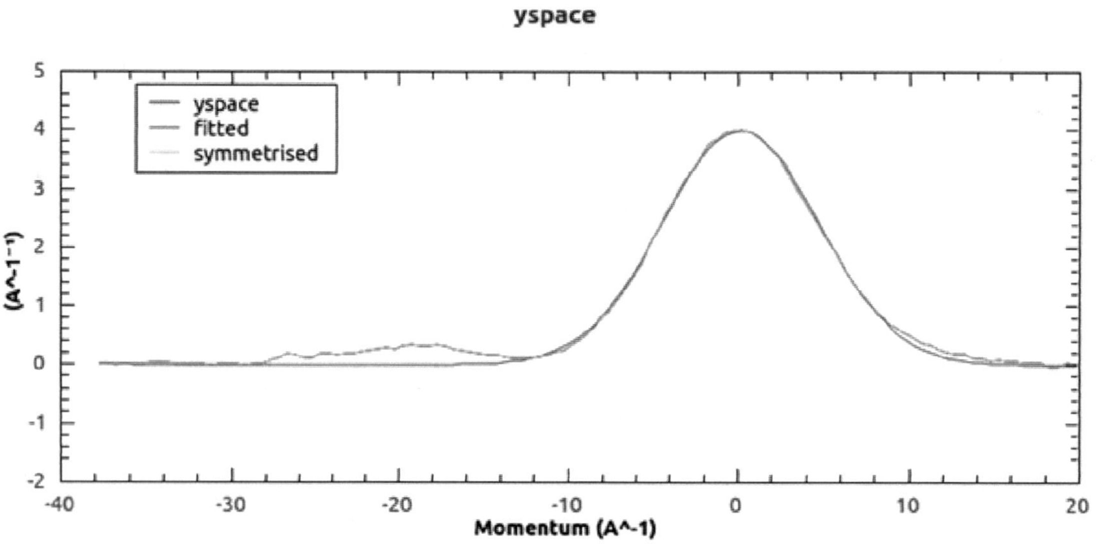

Figure 4. Plot of the time-of-flight data normalised and converted to y-space as the black line (behind the green). The red line shows the fit to the peak and the green line shows the symmetrised workspace which is in good agreement with y-space data.

5.6. Diffraction

Diffraction on VESUVIO is not yet fully implemented within MANTID. As the reduction of diffraction data is fairly trivial in comparison to the other requirements of VESUVIO data analysis, this can be handled by the existing *IndirectDiffractionReduction* routine that is used by other indirect geometry instruments. Preliminary trials with the existing implementation for other indirect instruments have shown that this should simply be a matter of adding the appropriate parameters for monitor thickness, attenuation, and area to VESUVIO's instrument parameter file.

5.7. Resolution Calculation

Calculation of the resolution of a particular mass in a sample can be performed using the calculate resolution function in *ncs.py*. This uses the same method to calculate the resolution that is used in the fit function described in Section 5.5. The output of running this function is a workspace with a single spectrum for the calculated resolution. Figure 5 shows a plot of the resolution for hydrogen and Table 8 shows the PYTHON code that generated the workspace shown in the plot.

5.8. Calibration of Instrument Parameters

The calibration routines for VESUVIO have been implemented as two MANTID algorithms following the procedures described originally in Ref. [44]. The first algorithm, called

Journal of Physics: Conference Series **571** (2014) 012009 doi:10.1088/1742-6596/571/1/012009

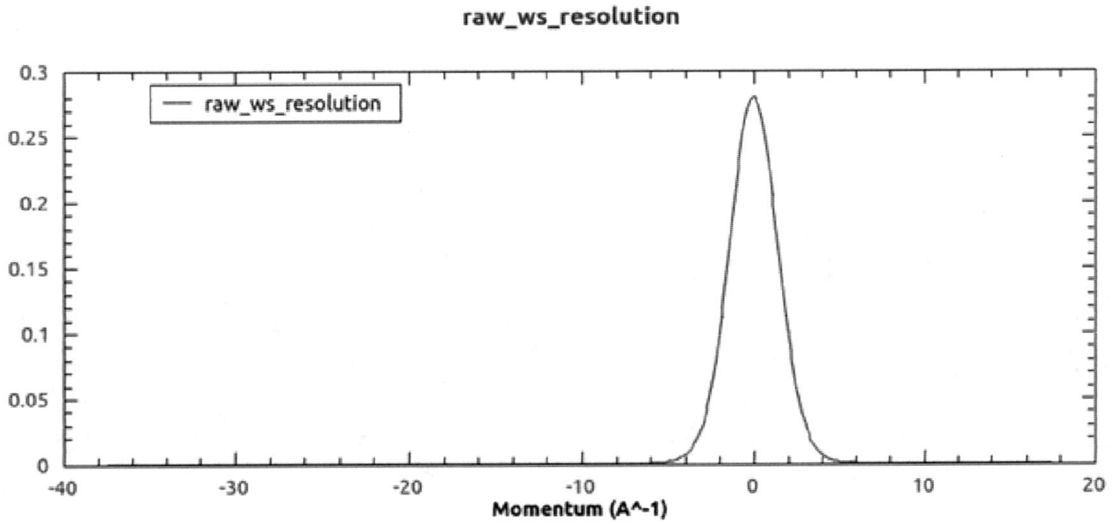

Figure 5. Plot of the resolution workspace generated using data in Table 8.

Table 8. Example script showing how to calculate the resolution for a particular mass.

```
1 import ncs
2 runs = "14188-14195"
3 spectra = "135-198"
4 diff_type="SingleDifference" # Allowed values=Single, Double, Thick
5 ip_file = "IP0004_10.par"
6 raw_ws = LoadVesuvio(Filename=runs, SpectrumList=spectra,
7 Mode=diff_type, InstrumentParFile=ip_file)
8 raw_ws = CropWorkspace(raw_ws, XMin=50.0, XMax=562.0)
9 mass = 1.0079
10 raw_ws = ConvertToYSpace(raw_ws, mass)
11 ncs.calculate_resolution(raw_ws, mass)
```

EVSCalibrationFit, is used to fit sample data in order to accurately obtain the values of the parameters for the instrument. A second MANTID algorithm built on top of the first is used to set up and run the fitting does the actual calculation of the instrument parameters called *EVSCalibrationAnalysis*. These two algorithms should only be run by instrument scientists and a calibration parameter file does not need to be recreated for every data analysis session.

The calibration fit program can either fit using a list of incident energies to calculate the expected centre point of a recoil peak in time-of-flight data or by taking a list of d-spacings and calculating the position of Bragg peaks. The algorithm also requires a file containing a set of reference parameters with which to calculate the expected peak positions. This is analogous to the procedures described in Ref. [7] where hand measured parameters were used as starting values and the fitting program used these to incrementally converge on the true value of the instrument parameters.

The analysis program makes use of the fit program to fit all of the parameters in the order defined by the original VESUVIO calibration paper [44], starting with the incident flight path

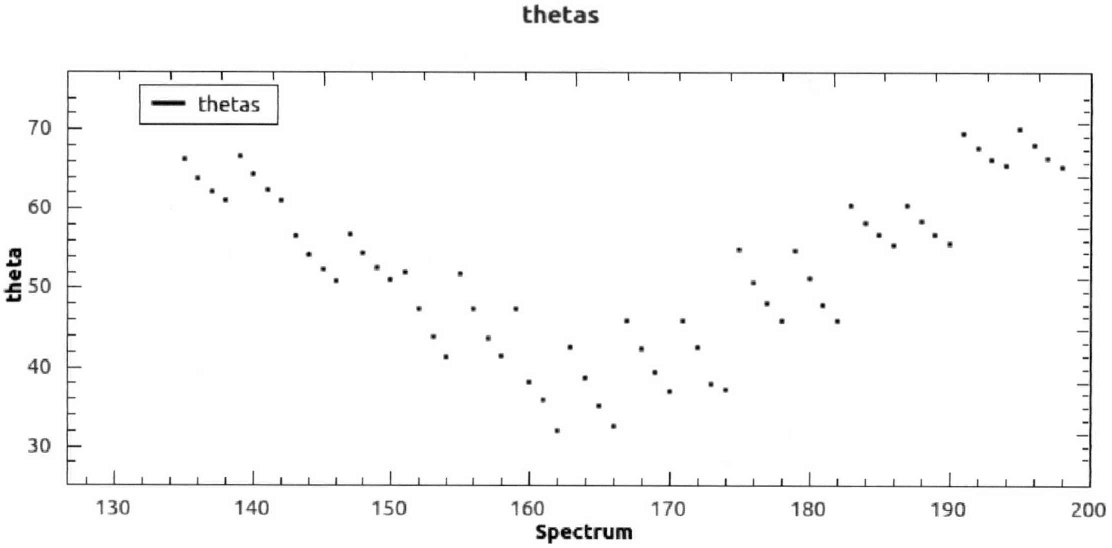

Figure 6. Plot showing the calculated scattering angle from a Pb sample.

and time delay using well defined uranium sample runs, then computing the values of the final energy and hence the final flight path and finally the scattering angles for all detectors (see Fig. 6).

5.9. GUI

VESUVIO does not currently have any GUI support within MANTID. Focus on development has been to get the underlying reduction and analysis routines working before focusing on usability. The current plan is for VESUVIO to have a completely new GUI listed under the indirect geometry section of MANTID. The current design would be to have a series of three tabs on the interface which would each deal with a separate part of reduction and analysis.

- *Loading:* This tab would broadly handle the procedures described in section 5.1 and would provide a user interface for using the *LoadVesuvio* algorithm to get raw data into MANTID as well as providing plotting functions for examining the captured time-of-flight data.
- *Corrections:* The corrections tab would provide a user interface for calculating both multiple scattering and gamma background corrections and applying them to data loaded in the first tab.
- *Fitting:* The fitting interface is the most complex in the series. This will provide support for setting up a fit using the same procedures used in the *ncs.py* module, but in a more user friendly manner than the current implementation achieves. This would provide support for both the Gaussian and Gram-Charlier functions as well as the ability to set the intensity constraints and Hermite polynomial expansion coefficients.

This is simply a loose plan based on previous discussions. The number and function of the tabs on the interface may change as development progresses. A proper design cannot be produced until the underlying framework is finalised. For example, there is a currently goal to support fitting a sample directly in y-space. It may be possible to integrate this with the fitting tab mentioned above, or depending on requirements, may be split to a fourth tab. Global fitting may well also require yet another separate interface.

Journal of Physics: Conference Series **571** (2014) 012009 doi:10.1088/1742-6596/571/1/012009

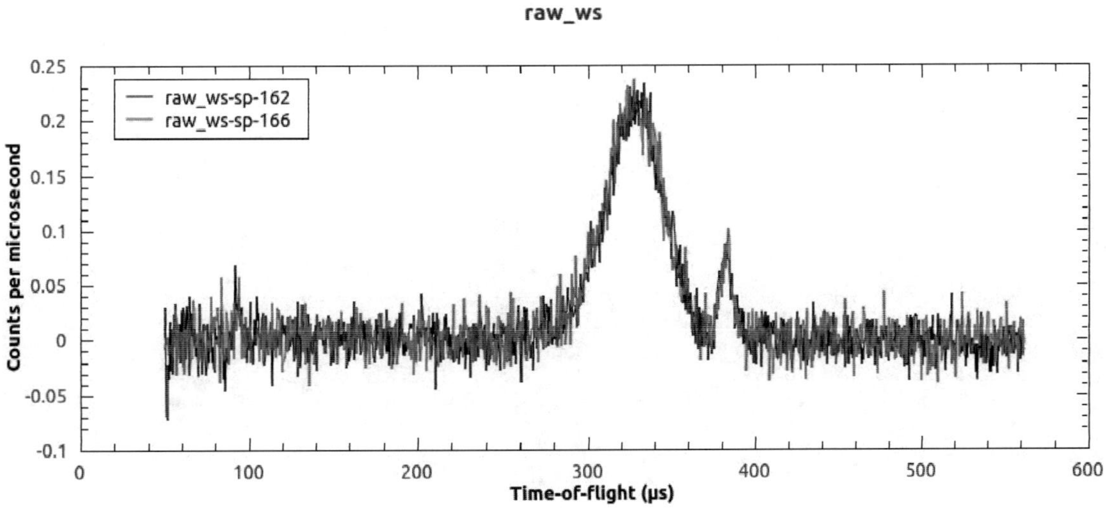

Figure 7. Plot of some raw time-of-flight data using the NCS plotting module using the script in 9. Checking the spectrum numbers in the key shows that the scattering angles of 33 and 32 degrees have been plotted.

5.10. Visualisation

VESUVIO analysis requires a few custom plotting commands to display data in ways which are not currently supported by the general MANTID framework. The final tab in the series would provide a user interface to allow the user to examine their data more easily with some helper functions which exist in the NCS plotting.py module. This is provides a single plot function with options to plot data by spectra, scattering angles, or detector banks.

This replicates the functionality for the various plotting routines of the old data analysis package [39]. Table 9 shows a basic example of plotting some spectra from a raw workspace by angle using the NCS plotting module. Figure 7 shows the resulting plot.

Table 9. Example PYTHON code showing how to plot spectra in a workspace within the scattering range of 30-35 degrees using the NCS plotting module.

```
import ncs_plotting
runs = "14188-14195"
spectra = "3-195"
diff_type="SingleDifference" # Allowed values=Single, Double, Thick
ip_file = "IP0004_10.par"
raw_ws = LoadVesuvio(Filename=runs, SpectrumList=spectra,
Mode=diff_type,InstrumentParFile=ip_file)
raw_ws = CropWorkspace(raw_ws, XMin=50.0, XMax=562.0)
ncs_plotting.plot(raw_ws, angles=(30,35))
```

Journal of Physics: Conference Series **571** (2014) 012009 doi:10.1088/1742-6596/571/1/012009

6. Conclusion

MANTID, the Manipulation and Analysis Toolkit for Instrument Data, is an open source and cross platform data analysis application specialising in neutron and muon scattering data. Recent development work undertaken to improve support for the indirect geometry spectrometer VESUVIO at ISIS, described in this work, demonstrates how MANTID can contribute to the success of neutron Compton scattering at providing unique insights into nuclear quantum dynamics of light nuclei in condensed matter and Mass-selective Neutron Spectroscopy of heavier nuclei important in the studies of functional materials.

7. Acknowledgments

The authors gratefully acknowledge the UK Science & Technology Facilities Council for financial support under Contract No. PR1200068 and access to beamtime at the ISIS Facility.

[1] Celli M, Zoppi M and Mayers J 1998 *Phys. Rev. B* **58** 242
[2] Bafile U, Celli M, Zoppi M and Mayers J 1998 *Phys. Rev. B* **58**(2) 791
[3] Zoppi M, Celli M and Soper A K 1998 *Phys. Rev. B* **58** 11905
[4] Pietropaolo A, Senesi R, Andreani C, Botti A, Ricci M and Bruni F 2008 *Phys. Rev. Lett.* **100** 127802
[5] Pantalei C, Pietropaolo A, Senesi R, Imberti S, Andreani C, Mayers J, Burnham C and Reiter G 2008 *Phys. Rev. Lett.* **100** 177801
[6] Feng Y, Ancona-Torres C, Rosenbaum T, Reiter G, Price D and Courtens E 2006 *Phys. Rev. Lett.* **97** 145501
[7] Senesi R, Pietropaolo A, Bocedi A, Pagnotta S and Bruni F 2007 *Phys. Rev. Lett.* **98** 138102
[8] Reiter G, Senesi R and Mayers J 2010 *Phys. Rev. Lett.* **105** 148101
[9] Mayers J, Burke T and Newport R 1994 *Journal of Physics-Condensed Matter* **6** 641
[10] Krzystyniak M, Adams M, Lovell A, Skipper N T, Bennington S M, Mayers J and Fernandez-Alonso F 2011 *Faraday Discuss.* **151** 171
[11] Mayers J 1993 *Phys. Rev. Lett.* **71** 1553
[12] Azuah R T, Stirling W G, Mayers J, Bailey I F and Sokol P E 1995 *Phys. Rev. B* **51** 6780
[13] Mayers J 1998 *Phys. Rev. Lett.* **80** 750
[14] Mayers J 2006 *Phys. Rev. B* **74** 014516
[15] Celli M, Zoppi M and Mayers J 1998 *Physical Review B* **58** 242
[16] Senesi R, Andreani C, Colognesi D, Cunsolo A and Nardone M 2001 *Phys. Rev. Lett* **86** 4584
[17] Peek D, Fujita I, Schmidt M and Simmons R 1992 *Physical Review B* **45** 9680
[18] Fradkin M, Zeng S X and Simmons R 1994 *Physical Review B* **49** 3197
[19] Fielding A, Timms D N and Mayers J 1998 *Europhysics Letters* **44** 255
[20] Abdul-Redah T, Georgiev P, Ross D, Krzystyniak M and Chatzidimitriou-Dreismann C 2005 *J. Alloys Compounds* **404** 787
[21] Dorner B 2005 *Journal of Neutron Research* **13** 267
[22] Krzystyniak M and Chatzidimitriou-Dreismann C 2005 *Journal of Physics: Condensed Matter* **20** 104249
[23] Andreani C, Colognesi D, Mayers J, Reiter G F and Senesi R 2005 *Adv. Phys.* **55** 377
[24] Seel A G, Sartbaeva A, Mayers J, Ramirez-Cuesta A J and Edwards P P 2011 *J. Chem. Phys.* **134** 114511
[25] Seel A G, Ceriotti M, Edwards P P and Mayers J 2012 *J. Phys.: Condens. Matter* **24** 365401
[26] Krzystyniak M and Abdul-Redah T 2010 *Phys. Rev. B* **82** 064301
[27] Mayers J and Reiter G 2012 *Measurement Science and Technology* **23** 045902
[28] Krzystyniak M, Richards S, Seel A and Fernandez-Alonso F 2013 *Phys. Rev. B* **88** 184304
[29] Arnold O, Bilheux J, Borreguero J, Buts A, Campbell S, Chapon L, Doucet M, Draper N, Leal R F, Gigg M, Lynch V, Markvardsen A, Mikkelson D, Mikkelson R, Miller R, Palmen K, Parker P, Passos G, Perring T, Peterson P, Ren S, Reuter M, Savici A, Taylor J, Taylor R, Tolchenov R, Zhou W and Zikovsky J 2014 *Nucl. Instrum. Meth. A* **764** 156
[30] Watson G I 1996 *J. Phys.: Condens. Matter* **8** 5955
[31] Sears V F 1984 *Phys. Rev. B* **30** 44
[32] West G B 1975 *Phys. Rep. C* **18** 263
[33] Mayers J, Andreani C and Baciocco G 1989 *Phys. Rev. B* **39** 2022
[34] Andreani C, Colognesi D and Pace E 1999 *Phys. Rev. B* **60** 10008
[35] Evans A C, Timms D N, Mayers J and Bennington S M 1996 *Phys. Rev. B* **53** 3023
[36] Mayers J and Abdul-Redah T 2004 *J. Phys.: Cond. Matter* **16** 4811
[37] Monserrat B, Drummond N D and Needs R J 2013 *Phys. Rev. B* **87** 144302

[38] https://github.com/mantidproject/scripts/tree/master/development/inelastic
[39] Mayers J 2011 User guide to vesuvio data analysis programs for powders and liquids Tech. Rep. RAL-TR-2011-003 Rutherford Appleton Laboratory, Chilton, UK URL http://purl.org/net/epubs/manifestation/5989
[40] Schooneveld M, Mayers J, Rhodes N, Pietropaolo A, Andreani C, Senesi R, Gorini G, Perelli-Cippo E and Tardocchi M 2006 *Review of scientific instruments* **77** 095103
[41] Seeger A, Taylor A D and Brugger R M 1985 *Nucl. Instrum. Meth. A* **240** 98
[42] Mayers J 2011 *Measurement Science and Technology* **22** 015903
[43] Mayers J and Adams M A 2011 *Nucl. Instrum. Meth. A* **625** 47
[44] Fielding A and Mayers J 2002 *Nucl. Instrum. Meth. A* **480** 680
[45] Andreani C, Colognesi D, Degiorgi E, Filabozzi A, Nardone M, Pace E, Pietropaolo A and Senesi R 2003 *Nucl. Instrum. Meth. A* **497** 535
[46] Pietropaolo A, Andreani C, Filabozzi A, Pace E and Senesi R 2007 *Nucl. Instrum. Meth. A* **570** 498
[47] Mayers J, Fielding A and Senesi R 2002 *Nucl. Instrum. Meth. A* **454** 481

Journal of Physics: Conference Series **571** (2014) 012010 doi:10.1088/1742-6596/571/1/012010

Discussion: Measurement and Instrumentation

C. Andreani,[1,2] M. Ceriotti,[3] F. Fernandez-Alonso,[4,5] G. Gorini,[6] M. Krzystyniak,[4,7] A. Parmentier,[1] S. Parker,[4] A. Seel,[4,8] R. Senesi,[1,2] L. Walewski,[9]

[1] Università degli Studi di Roma Tor Vegata, Dipartimento di Fisica and Centro NAST, via della Ricerca Scientifica 1, Rome 00133, Italy
[2] CNR-IPCF Sezione di Messina, Messina, Italy
[3] Laboratory of Computational Science and Modelling, NQ Institute of Materials, École Polytechnique Fédérale de Lausanne, CH-1015 Lausanne, Switzerland
[4] ISIS Facility, Rutherford Appleton Laboratory, Chilton, Didcot, Oxfordshire OX11 0QX, United Kingdom
[5] Department of Physics and Astronomy, University College London, Gower Street, London WC1E 6BT, United Kingdom
[6] Consiglio Nazionale della Ricerche-IPCF, Messina, Italy
[7] School of Science and Technology, Nottingham Trent University, Clifton Campus, Nottingham NG11 8NS, United Kingdom
[8] Department of Chemistry, University of Oxford, South Park Road, Oxford OX1 3QR, United Kingdom
[9] Lehrstuhl für Theoretische Chemie, Ruhr-Universität Bochum, 44780 Bochum, Germany

1. Epithermal Neutron Instrumentation at ISIS (Giuseppe Gorini, Milano, Italy)

Dr Parker opened the discussion of the lecture by Prof Gorini: What are the lightest elements that NRT/NRCA can measure?
Prof Gorini replied: We have compiled a periodic table showing good, bad and 'no' elements, according to their detection sensitivity of the techniques.

Dr Krzystyniak communicated: Do you have plans for thermography studies using NRT/NRCA?
Prof Gorini communicated in response: There are no plans for T measurements at present, but other groups in Japan are exploring the feasibility.

2. VESUVIO Now/VESUVIO When? (Andrew Seel, STFC, UK)

Dr Ceriotti opened the discussion of the lecture by Dr Seel: It seems that there are different approaches around for performing the fitting of the $n(p)$ to the NCS data: 1. Fit the raw data from individual detectors and average the kinetic energy (and/or other fit parameters) obtained from individual channels; 2. Average groups of detectors with similar properties and fit the averaged data; 3. Fit to all the individual channels simultaneously. What are the differences of

Journal of Physics: Conference Series **571** (2014) 012010 doi:10.1088/1742-6596/571/1/012010

these approaches?
Dr Seel replied: Individual, detector-by-detector fits can be thought of as the traditional way of analysing data from VESUVIO, since the detector geometry determines the transform of mass specific neutron Compton profiles from y-space to time-of-flight. In fact the initial tests on the reliability of y-scaling for subsequent analysis was dependant upon this approach. This approach is limited by the levels of noise in VESUVIO spectra (the instrument is flux limited), as is a global fit of the same $J(y)$ to each detector simultaneously. However, for the case of heavier masses, the recoil position and peak widths vary little across small detector ranges, allowing for the possibility to direct sum spectra in time-of-flight, and fit to J(y) transformed using an average of the detector geometries. This would enable more stable fits across the instrument, effectively reducing the noise of the measured spectra.

Dr Seel further communicated in response: VESUVIO is limited principally by flux, necessitating long counting times (of the order of 15-24 hours). An improvement of even a factor of two through the upgrade of the ISIS target-station would revolutionise the way science is conducted on VESUVIO.

Prof Fernandez-Alonso commented: Such upgrades could provide possible improvement factors of 10 or even 100.

Dr Walewski queried: Why it is not possible to get constant q scans on VESUVIO?
Dr Seel replied: This is principally a geometry problem, bearing in mind we operate in inverted geometry and need to cycle secondary analyser foils in the scattered beam for each detector. The use of y-scaling overcomes the issue of constant q, and allows us to probe measurable properties such as momentum distributions and mean kinetic energies of interest to the community.

3. Direct and Inverse Geometry Instruments for Proton Momentum Distribution Studies (Roberto Senesi, Rome, Italy)

Prof Andreani opened the discussion of the lecture by Dr Senesi: Why not using the positive y-side of NCP from SEQUOIA to compare with the positive one from VESUVIO?
Dr Senesi replied: This is a good suggestion, and this test will be carried out.

Dr Seel communicated: What is the magnitude of FSEs on SEQUOIA, compared to VESUVIO?
Dr Senesi communicated in response: At wave vector values between 20 and 30 Å^{-1}, we were able to detect only the $\frac{1}{q}$-term.

Dr Krzystyniak commented: The y-range accessed on VESUVIO is limited on positive high y-values, while the opposite is true for SEQUOIA.

Journal of Physics: Conference Series **571** (2014) 012011 doi:10.1088/1742-6596/571/1/012011

Ab initio simulation of particle momentum distributions in high-pressure water

M. Ceriotti[1]

[1]Laboratory of Computational Science and Modelling, NQ Institute of Materials, École Polytechnique Fédérale de Lausanne (EPFL), CH-1015 Lausanne, Switzerland

E-mail: `michele.ceriotti@epfl.ch`

Abstract.
Applying pressure to water reduces the average oxygen-oxygen distance, and facilitates the delocalisation of protons along the hydrogen bond. This pressure-induced delocalisation is further enhanced by the quantum nature of hydrogen nuclei, which is very significant even well above room temperature. Here we will evaluate the quantum kinetic energy and the particle momentum distribution of hydrogen and oxygen nuclei in water at extreme pressure, using ab initio path integral molecular dynamics. We will show that (transient) dissociation of water molecules induce measurable changes in the kinetic energy hydrogen atoms, although current deep inelastic scattering experiments are probably unable to capture the heterogeneity of the sample.

Treating the nuclei as classical particles is an approximation which is widely made in atomistic computer simulations, even in cases in which the electronic degrees of freedom are treated quantum mechanically by first-principles methods. Whenever light nuclei such as hydrogen are present, however, their quantum nature will have a sizable effect on physical observables even well above room temperature, that must be considered if one wants to capture quantitatively the properties of the system.

A typical example in which nuclear quantum effects (NQEs) cannot be neglected is the evaluation of the particle momentum distribution of light nuclei, that deviates dramatically from the Maxwell-Boltzmann predictions and that can be measured in deep inelastic neutron scattering (DINS) experiments [1]. The standard method to describe the quantum nature of nuclei in simulations is based on the path integral formalism [2–5], which however involves a very significant computational overhead compared to a classical simulation, making it very challenging to include NQEs in *ab initio* simulations. Recently it has been shown that correlated-noise Langevin dynamics can be used to model NQEs inexpensively [6, 7], and that a combination of path integral molecular dynamics and colored noise (PIGLET) dramatically reduces the cost of attaining quantitative accuracy [8, 9].

Here, such a combination of path integral methods and generalized Langevin dynamics will be used to evaluate the particle momentum distribution (PMD) of oxygen and hydrogen nuclei in water at 10GPa and 750K. In this region of the phase diagram the mean O–O distance is considerably reduced relative to ambient temperature and pressure, increasing the probability of water dissociation to a point at which a significant fraction of auto-ionisation events can be observed within the limited time accessible to first-principles simulations [10]. Quantum effects facilitate dissociation even further [11], and it is interesting to evaluate the impact of

Journal of Physics: Conference Series **571** (2014) 012011 doi:10.1088/1742-6596/571/1/012011

autoionisation on the hydrogen and oxygen mean kinetic energy. After having reported the estimates that can be obtained by computer simulations, it will be discussed whether and how it may be possible to measure these effects by DINS experiments.

1. Computational details

Simulations were performed using i-PI, a Python wrapper that makes it possible to do path integral and PIGLET simulations using any electronic structure code as a client to compute the Born-Oppenheimer energy and forces [11]. In this case, CP2K was used as the back-end [12, 13].

The supercell contained 64 water molecules, equilibrated at the desired thermodynamic conditions using an empirical forcefield [14]. Then, 25 ps of NVT equilibration with classical nuclei were perfomed with *ab initio* forces, using the BLYP exchange-correlation functional [15, 16] and GTH pseudopotentials [17]. Wave functions were expanded in the Gaussian DZVP basis set, while the electronic density was represented using an auxiliary plane wave basis, with a kinetic energy cutoff of 300 Ry.

Then, constant-pressure production simulations were performed using the barostatting procedure described in [11], using a more converged TZV2P basis set, a density cutoff of 800 Ry, and a plane wave grid consistent with a cubic reference cell with a side of 10.5 Å (the higher cutoff was necessary to converge the stress tensor). Empirical Van der Waals corrections, which are essential to obtain a reasonable description of water at constant pressure [18], were included based on the D3 scheme [19–21]. The quantum nature of nuclei was modelled using a 4-bead ring polymer, supplemented with PIGLET colored noise. The use of colored noise makes it possible to converge simulation of water at room temperature using as little as six path integral replicas [9], and so one can expect that at this much higher temperature a slightly reduced number of replicas will be sufficient for convergence of quantum properties. The cell momentum was thermostatted using an optimal-sampling, classical GLE thermostat. A time step of 0.5fs was used, and a total of 50ps of PIGLET dynamics were performed to accumulate statistical averages.

NQEs enhance greatly the delocalisation of protons along the hydrogen bond, so that even in room-temperature neutral water there is a sizeable probability of observing transient autoionization events where a hydrogen atom momentarily visits configurations in which it is closer to the acceptor oxygen atom than to the oxygen it is covalently bound to [22]. This makes it somewhat problematic to identify charged H_3O^+ and OH^- species based only on structural information, so here we use a simple-minded approach to identify configurations that are significantly distorted relative to a neutral water molecule environment, without trying to distinguish between transient excursions and genuine autoionization.

Following Refs. [10, 11], we attributed to each oxygen atom the hydrogens that are within its Voronoi polyhedron (i.e. those that are closest to it than to any other oxygen), and labelled as neutral species the oxygen atoms with two hydrogen partners, positive '+' species the oxygen atoms to which three H have been assigned, and '-' species the oxygen atoms to which a single proton belongs. We also refer to H^+ as the hydrogens that are assigned to a O^+, H^- as the hydrogens that are assigned to a O^-, and to just H as the protons that are attached to a 'neutral' oxygen. According to this definition, simulations using PIGLET contain about 1% of O^+ and O^- species, almost 20 times as much as simulations in the same conditions that ignore NQEs [11].

2. Quantum kinetic energies and particle momentum distributions

Atomistic computer simulations make it possible to interrogate with great detail the properties of the system being studied. For instance, one can compute conditional averages that evaluate physical observables for only a subset of the atoms. In particular, one can ask what is the kinetic energy of hydrogen and oxygen nuclei as a function of the formal charge that has been assigned based on the procedure outlined above. What is more, one can evaluate for each of the species the transient anisotropic Gaussian (TAG) approximation of the particle momentum distribution [9], giving even further insight into the difference between environments.

The TAG approximation evaluates an approximant of the kinetic energy tensor for a particle at a given time by performing a moving average of the centroid-virial kinetic tensor estimator $T_{\alpha\beta}$ over a triangular window of width Δt:

$$\begin{aligned} T_{\alpha\beta} =& \frac{\delta_{\alpha\beta}}{2\beta} + \frac{1}{4P}\sum_i \left[(q_{i\alpha} - \bar{q}_\alpha)\frac{\partial V}{\partial q_{i\beta}} + (q_{i\beta} - \bar{q}_\beta)\frac{\partial V}{\partial q_{i\alpha}}\right]; \\ T_{\alpha\beta}(t;\Delta t) =& \frac{1}{\Delta t}\int_{t-\Delta t}^{t+\Delta t} T_{\alpha\beta}(t')\left(1 - \frac{|t-t'|}{\Delta t}\right) \mathrm{d}t'. \end{aligned} \tag{1}$$

Here, P is the number of path integral replicas, $q_{i\alpha}$ identifies the α Cartesian coordinate of the tagged particle in the i-th replica, $\bar{\mathbf{q}}$ is the position of the centroid of the particle, V is the physical potential and β the inverse temperature. One can compute the three principal components of $\mathbf{T}(t;\Delta t)$, E_α, and average them over the course of the trajectory. This procedure gives only an approximation of the three components of the mean kinetic energy tensor: a too short window means that the average picks up an instantaneous value of the centroid virial estimator (which is not physically meaningful), while a too large window would average over different orientations of the molecule and tend towards giving an isotropic tensor. By conditionally averaging over particles that are in a given formal charge state at time t one can evaluate a TAG approximation to the kinetic energy resolved between O^+, O, O^- and H^+, H, H^-.

Figure 1. Total mean kinetic energy of hydrogen (top) and oxygen (bottom) atoms. Moving averages are performed with a time window of amplitude Δt, and conditionally based on the charge state of each atom at the centre of the moving window. O^+, O, O^- and H^+, H, H^- species are indicated by different symbols, as indicated in the legend.

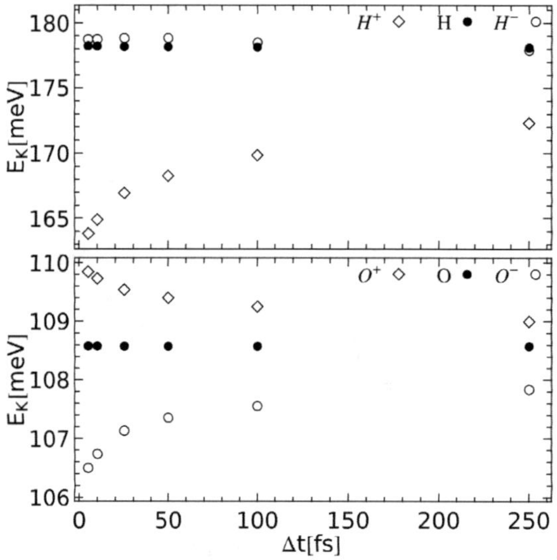

One should stress that using the TAG approximation in this context is stretching considerably its range of applicability. For one thing, since the charged defects have a fluxional, rapidly changing nature, when averaging over larger Δt one often picks up configurations that are not in the same formal charge state as the central frame that determines the conditional averaging. This means that as Δt increases the *total* mean kinetic energies (that are not affected by the TAG procedure if the average is performed over all atoms of the same element) of the different charged species become less and less different from each other (see Figure 1).

Figure 2. Principal components of the mean kinetic energy of hydrogen (top) and oxygen (bottom) atoms, as computed within the TAG approximation with a time window of amplitude Δt. Averages are performed conditionally based on the charge state of each atom at the centre of the moving window. O^+, O, O^- and H^+, H, H^- species are indicated by different symbols, as indicated in the legend, and the three components of the kinetic energy tensor are drawn in red, green and blue.

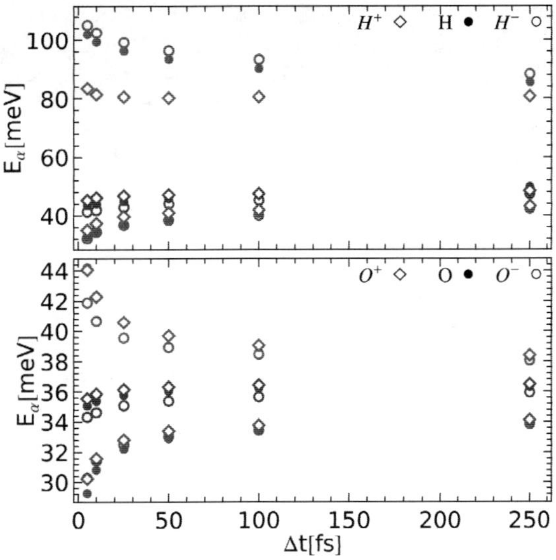

Furthermore, at 750K the orientational relaxation of water is much faster than at room temperature, and there is no obvious plateau between the Δt that is needed to converge the mean centroid virial estimator and the Δt at which the rotation of water molecules averages out the instantaneous environment felt by protons (see Figure 2).

Nevertheless, use of the TAG approximation is necessary here, as the alternative approach of defining molecular axes along which to perform a kinetic energy decomposition is not viable for the charged species, and since an open path calculation would be exceedingly hard to converge, given that the approximation of opening one path per molecule [23] is hard to justify when there is a significant degree of proton exchange between molecules. Let us take rather arbitrarily Δt = 50fs as the averaging window for which we will discuss results. In the absence of a clear plateau, this seems a reasonable compromise between the conflicting goals of averaging the unphysical instantaneous values of the centroid virial estimator and the need to capture a rapidly changing instantaneous environment.

A first observation is that even at 750K atoms in water are far from a classical behaviour. At this temperature the classical mean kinetic energy per atom would have a value of 97 meV. Oxygen atoms have 10% higher E_K, and hydrogen atoms have nearly double E_K compared to the classical predictions. This is mostly due to the zero-point energy in the stretching mode. Differences between charged species is noticeable but small, amounting to a few meV for most components. The only large difference is observed for H^+ species, that have a weakened stretching mode resulting in a drop of more than 15 meV in the largest principal component of the kinetic energy tensor. As it is often the case in water [24–26], there is a degree of cancellation between the changes of the different components of the kinetic energy, so that the overall kinetic energy change between H and H^+ is reduced to about 11 meV when evaluated with the same moving window of 50fs. H^- does not show large differences to "neutral" H atoms, with the components E_α changing by less than 3 meV, and a near-perfect cancellation when the difference in the total E_K is considered.

Note that due to the low concentration of charged species, the kinetic energy components averaged over *all* the H or O atoms is almost indistinguishable from the E_αs for the dominant neutral H and O species. This points at an inherent limitation of DINS experiments, that shall be discussed in more detail in the next section.

Figure 3. Particle momentum distribution approximated as the spherical average of a multivariate Gaussian, with the principal components obtained from the TAG approximation with Δt =50fs. The three lines correspond to H^+, H, H^- species, as indicated in the legend.

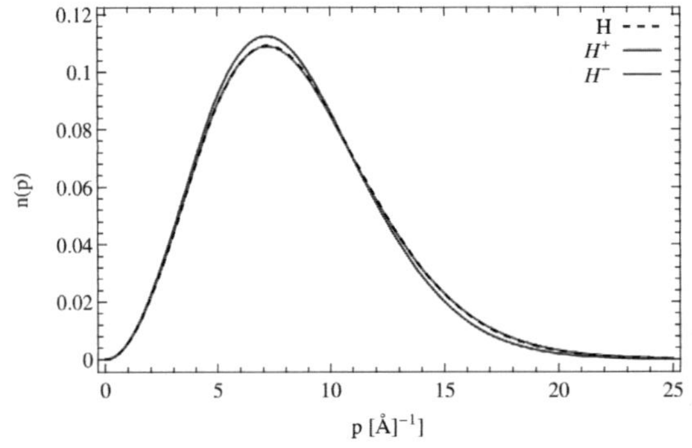

3. Detecting heterogeneity in DINS experiments

Neutron Compton scattering is perhaps the most direct probe of the quantum nature of nuclei in the condensed phase, fully accounting for anharmonicity [27], and having been used to probe the environment of protons in a variety of different materials [28–30]. In the present case it is clear that the system – albeit homogeneous on a macroscopic scale – contains a variety of different local environments, that in the specific case of the charged H^+ species imply a sizeable change in the kinetic energy tensor compared to the H species.

Fig. 3 shows that the particle momentum distribution of H^+, taken separately, would differ significantly from that of the "neutral" hydrogen atoms, by an amount sufficient to be unambiguosly detected by DINS. As we discussed before, the small concentration of charged species means that the signal would be probably too weak to be detected against that of the majority species. This lead us to ask ourselves, more in general, whether DINS can provide information on the heterogeneity of a material, a question that is relevant to the interpretation of experimental results and to the comparison with simulations.

Let us consider for example the most favourable case of a hypotetical sample containing 50% of H^- and 50% of H^+ species. Let us assume that the overall PMD arising from the sample will then be the arithmetic mean of the multivariate-Gaussian TAG approximant of the two species as obtained in our simulations. Figure 4 shows that one can obtain a near-perfect fit (undistinguishable within the current experimental uncertainty) to this average distribution using a *single* multivariate Gaussian model. This observation highlights one problem of DINS experiments – namely that spherical averaging washes away much of the information content of

Figure 4. The top panel shows the mean between the multivariate-Gaussian particle momentum distributions of H^+ and H^- species, and the distribution obtained by fitting a single multivariate Gaussian model to the former curve. The relative discrepancy between the two curves is also plotted in the lower panel.

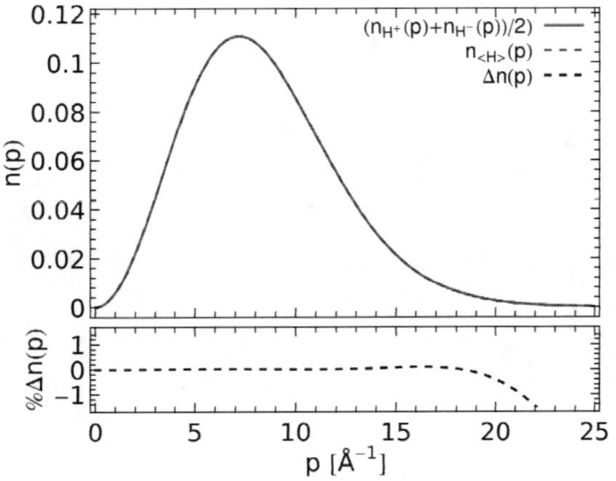

the particle momentum distribution, so that it is almost impossible to single out contributions from different environments in the sample, and just a "mean field" description can be obtained.

4. Outlook

The present work shows that the quantum nature of light nuclei is still very pronounced even much above room temperature, and that the quantum kinetic energy of particles varies significantly between nuclei that are experiencing different environments. At the same time, it highlights limitations of the current approaches to simulate and measure NQEs in the condensed phase. From the modelling point of view, it shows the limits of a recently-developed approximation to estimate a Gaussian model of the particle momentum distribution based on an "instantaneous" kinetic energy tensor. While this TAG approximation makes it possible to obtain the PMD inexpensively, it can only provide qualitative information, so an approach to make conventional open path simulations more affordable is much needed.

On the experimental side, this paper points at a significant limitation of current DINS techniques – namely the fact that it is practically impossible to extract information on the heterogeneity of the sample, as the PMD arising from a very inhomogeneous sample would be undistinguishable from that of a homogeneous system made up of an effective, mean environment. A possible approach that we can suggest to extract some site-specific information would be to exploit the different propensity of hydrogen isotopes to occupy different sites, and to perform measurements with different isotope tagging. In systems where proton exchange is minimal, this would allow one to precisely measure the PMD of specific sites. In systems such as water in which proton mixing happens much more rapidly than a measurement can be performed, it might still be possible to infer some insight on the heterogeneity of the material by comparing measurements performed at different isotope concentrations.

Acknowledgments

We thank Greg Voth for suggesting to analyse the output of these simulations to compute the particle momentum distribution, David Manolopoulos for suggestions and insightful discussion, and acknowledge generous allocations of computer time from the Swiss National Supercomputing

Journal of Physics: Conference Series **571** (2014) 012011 doi:10.1088/1742-6596/571/1/012011

Centre (project ID s338 and s466).

References

[1] Andreani C, Colognesi D, Mayers J, Reiter G F and Senesi R 2005 *Adv. Phys.* **54** 377–469
[2] Feynman R P and Hibbs A R 1964 *Quantum Mechanics and Path Integrals* (New York: McGraw-Hill)
[3] Chandler D and Wolynes P G 1981 *J. Chem. Phys.* **74** 4078–4095
[4] Parrinello M and Rahman A 1984 *J. Chem. Phys.* **80** 860
[5] Ceperley D M 1995 *Rev. Mod. Phys.* **67** 279–355
[6] Ceriotti M, Bussi G and Parrinello M 2009 *Phys. Rev. Lett.* **103** 30603
[7] Dammak H, Chalopin Y, Laroche M, Hayoun M and Greffet J J 2009 *Phys. Rev. Lett.* **103** 190601
[8] Ceriotti M, Manolopoulos D E and Parrinello M 2011 *J. Chem. Phys.* **134** 84104
[9] Ceriotti M and Manolopoulos D E 2012 *Phys. Rev. Lett.* **109** 100604
[10] Schwegler E, Galli G, Gygi F and Hood R 2001 *Phys. Rev. Lett.* **87** 265501
[11] Ceriotti M, More J and Manolopoulos D E 2014 *Comput. Phys. Commun.* **185** 1019–1026
[12] CP2K http://www.cp2k.org
[13] VandeVondele J, Krack M, Mohamed F, Parrinello M, Chassaing T and Hutter J 2005 *Comput. Phys. Commun.* **167** 103–128
[14] Habershon S, Markland T E and Manolopoulos D E 2009 *J. Chem. Phys.* **131** 24501
[15] Becke A D 1988 *Phys. Rev. A* **38** 3098
[16] Lee C, Yang W and Parr R G 1988 *Phys. Rev. B* **37** 785
[17] Goedecker S, Teter M and Hutter J 1996 *Phys. Rev. B* **54** 1703–1710
[18] Schmidt J, VandeVondele J, Kuo I F W, Sebastiani D, Siepmann J I, Hutter J and Mundy C J 2009 *J. Phys. Chem. B* **113** 11959–64
[19] Grimme S, Antony J, Ehrlich S and Krieg H 2010 *J. Chem. Phys.* **132** 154104
[20] Guidon M, Hutter J and VandeVondele J 2009 *J. Chem. Theory Comput.* **5** 3010–3021
[21] Guidon M, Hutter J and VandeVondele J 2010 *J. Chem. Theory Comput.* **6** 2348–2364
[22] Ceriotti M, Cuny J, Parrinello M and Manolopoulos D E 2013 *Proc. Natl. Acad. Sci. USA U. S. A.* **110** 15591–6
[23] Morrone J A, Srinivasan V, Sebastiani D and Car R 2007 *J. Chem. Phys.* **126** 234504
[24] Li X Z, Walker B and Michaelides A 2011 *Proc. Natl. Acad. Sci. USA* **108** 6369–6373
[25] Markland T E and Berne B J 2012 *Proc. Natl. Acad. Sci. USA* **109** 7988–7991
[26] Romanelli G, Ceriotti M, Manolopoulos D E, Pantalei C, Senesi R and Andreani C 2013 *J. Phys. Chem. Lett.* **4** 3251–3256
[27] Mayers J, Reiter G F and Platzman P 2002 *J. Mol. Struct.* **615** 275–282
[28] Andreani C, Degiorgi E, Senesi R, Cilloco F, Colognesi D, Mayers J, Nardone M and Pace E 2001 *J. Chem. Phys.* **114** 387–398
[29] Reiter G F, Senesi R and Mayers J 2010 *Phys. Rev. Lett.* **105** 148101
[30] Reiter G F, Kolesnikov a I, Paddison S J, Platzman P M, Moravsky a P, Adams M a and Mayers J 2012 *Phys. Rev. B* **85** 045403

Journal of Physics: Conference Series **571** (2014) 012012 doi:10.1088/1742-6596/571/1/012012

Nuclear quantum effects in a 1-D model of hydrogen bonded ferroelectrics

K T Wikfeldt[1,2]

[1]Science Institute, University of Iceland, Iceland, and [2]Nordita, 10691 Stockholm, Sweden

E-mail: wikfeldt@hi.is

Abstract. A one dimensional model of a coupled hydrogen (H) bonding chain is developed and parametrized to density functional theory (DFT) calculations on squaric acid, a prototypical H-bonded antiferroelectric crystal. The energetics of single and collective proton jumps and its dependence on H-bond length, as obtained by DFT, is reproduced quite well in the model despite its simplicity where only hydrogen and oxygen atom positions in (O-H...O) H-bonds and nearest-neighbor coupling between H positions are explicitly included. Classical and path-integral molecular dynamics simulations are performed to shed light on nuclear quantum effects and how they influence the paraelectric phase transition. A large H/D isotope shift in the transition temperature T_C as well as a geometric isotope effect is obtained in good agreement with experiment. Fixing the O-O bond length results in shifts of T_C to higher temperature but a pronounced isotope shift of T_C remains, highlighting the importance of quantum effects beyond the geometrical changes in H-bonds accompanying isotopic substitution. Intermediate between fully atomistic models and simpler Ising-type models, the proposed H-bond chain model is a useful toy model for investigating microscopic mechanisms behind phase transitions in H-bonded ferroelectrics and the detailed role of quantum fluctuations.

1. Introduction

One dimensional (1D) chains of molecules connected by hydrogen (H) bonds play a key role in a range of important phenomena in nature. Bridges consisting of several water molecules are fundamental to proton transfer processes in enzyme catalysis [1] and in membrane water channels [2] and the mechanisms of such concerted proton transfers are thus of great biological interest [3]. Water molecules confined inside carbon nanotubes have been investigated as an ideal model system of proton conduction along water wires [4, 5], and the presence of tunneling-assisted concerted proton jumps between molecules in cyclic water hexamers in ice has recently been demonstrated both experimentally [6] and theoretically [7].

Directional H-bond chains also lead to ferroelectric properties in a range of H-bonded crystals [8, 9]. These H-bonded ferroelectrics have been actively investigated since the discovery of ferroelectricity in rochelle salt in 1921 [10], both due to their many possible uses in technological applications [11] and because their fundamental properties are unusual and not yet fully understood. In particular the prototypical KDP system, composed of H-bonded KH_2PO_4 molecules, has been intensively investigated [12, 13]. Organic H-bonded ferroelectrics constitute another class of materials that have been proposed [14] as flexible low-cost, environmentally friendly alternative components of electronic devices, especially after the discovery of room-temperature ferrolectricity in the croconic acid crystal [15, 16, 17]. Croconic acid is a cyclic oxocarbon ($C_5O_5H_2$) in which two inequivalent chains of H-bonds determine the overall polarization. A related but simpler organic ferroelectric is squaric acid, H_2SQ, which is composed of stacked planar sheets of $C_4O_4H_2$ molecules with opposite polarization in neighboring planes, leading to antiferroelectric properties. Both experimental [18] and theoretical [19] studies have shown that the two orthogonal interpenetrating H-bonding chains in H_2SQ are effectively decoupled, suggesting that a 1D picture may be able to capture fundamental properties of this system.

A key aspect of most H-bonded ferroelectrics that determines their ferroelectric behavior is the motion of H atoms in correlated double well potentials. This correlation consists in H atom positions in neighboring H-bonds being strongly coupled due to the energetic requirement of satisfying the ice rules: in the case of squaric acid each molecule participates in 4 H-bonds, two donating and two accepting. An intriguing property of these systems is the giant isotope shift on T_C. In squaric acid T_C shifts from 373 K in the hydrogenated crystal to 520 K in the deuterated crystal. An early explanation for this behavior was based on quantum tunneling of protons in double-well potentials as proposed by Blinc [20]. After the experimental observation of large geometrical isotope effects (often called Ubbelohde effects [21]), where O-O bond lengths increase significantly upon deuteration, it was argued that quantum tunneling is not necessary to explain this large shift of T_C since H and D atoms experience different double-well potentials that lead to different values of T_C [22]. However, more recent work has again emphasized the involvement of tunneling in H-bonded ferroelectrics. An experimental measurement of the proton momentum distribution in KDP indicated the presence of tunneling above T_C [23], and a first-principles computational investigation suggested tunneling in clusters of correlated protons and heavy atom displacements as the underlying mechanism behind the geometric isotope effect and shift of T_C [13, 24]. A comprehensive understanding of H-bonded ferroelectrics, in particular the role of nuclear quantum effects, is thus yet to be developed.

In recent work we used ab initio path integral molecular dynamics (PIMD) simulations, employing density functional theory (DFT), to investigate nuclear quantum effects on the structure and ordering behavior of squaric acid [25]. We first compared the proton transfer barrier obtained by various density functionals to more accurate methods, *i.e.* the explicitly correlated MP2 and RPA methods, and found that vdW-DF2 [26] gave the best agreement although it slightly underestimated the barrier. The vdW-DF2 functional was subsequently used in PIMD simulations, revealing a very large influence of nuclear quantum effects. Concerted proton jumps along H-bonding chains were found to be facilitated by synchronous quantum tunneling of several protons. Although accurate in terms of the full-dimensional quantum mechanical description of both electronic and nuclear degrees of freedom, this previous study was limited to a small system size and short simulation times due to the significant computational cost. Simpler models are required to investigate in more detail the paraelectric phase transition, cooperative effects along H-bonding chains and the influence of nuclear quantum effects.

In this work a simple model for a coupled 1D H-bonding chain is presented. It has been parametrized to DFT calculations on squaric acid, a system where a 1D description can be expected to be a good approximation. The structure of the planar building blocks of squaric acid is shown in Fig. 1(a) while the H-bond chain model is illustrated in Fig. 1(b). For more complex materials the model can be generalized to higher dimensions but one might expect that physical insights derived here will be useful for other similar systems.

Extensive molecular dynamics (MD) simulations have been performed on the model to investigate its disordering phase transition, and PIMD simulations have been used to study nuclear quantum effects. In the next section details of the model and the DFT calculations used to parametrize it will be described. Results will be discussed in section 3 and conclusions summarized in section 4.

Journal of Physics: Conference Series **571** (2014) 012012 doi:10.1088/1742-6596/571/1/012012

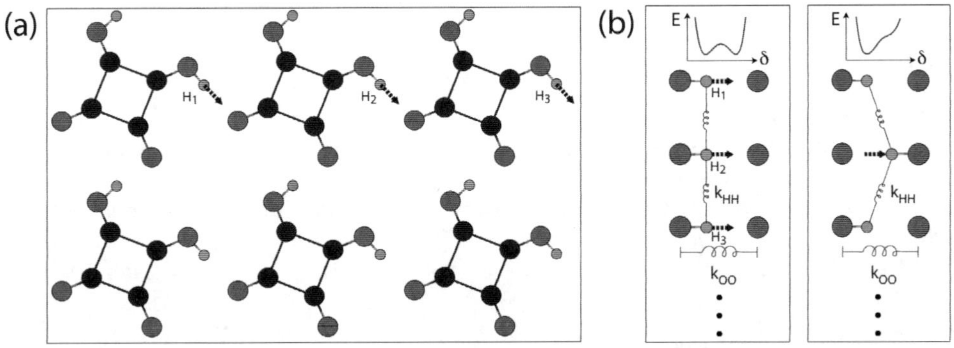

Figure 1. (a) Crystal structure in one plane of squaric acid showing a 3×2 supercell. The primitive unit cell is composed of two molecules in neighboring parallel planes. A collective transfer of three protons (H_1, H_2 and H_3) along one H-bonding chain is shown by dashed arrows. (b) Schematic of the 1D H-bond chain model. Protons in neighboring H-bonds (illustrated vertically) are harmonically coupled by springs k_{HH}, the bond length R is coupled by springs k_{OO} around an equilibrium value R_{eq}, and the protons experience a potential that depends on the distance R (see text). The left panel shows the equilibrium structure and (top) a double-well potential for a synchronous jump of all protons, while the right panel shows the jump of one proton across the H-bond and (top) the corresponding single-well potential.

2. Methods

2.1. The HBC model

The proposed H-bond chain (HBC) model contains three terms in the Hamiltonian: (i) an adiabatic ground state energy surface based on two diabatic states suggested by McKenzie [27] for the motion of protons in H-bonds, (ii) a harmonic potential for the oxygen-oxygen distance and (iii) a harmonic potential between protons in nearest-neighbor H-bonds. The Hamiltonian for N H-bonds is thus written as

$$\mathcal{H} = \sum_{i=1}^{N} (V(r_i, R_i) + V_{OO}(R_i) + V_{HH}(r_i, r_{i+1})) \tag{1}$$

where r and R denote hydrogen position and oxygen-oxygen distance, respectively (to be denoted by d_{OH} and d_{OO} in following sections). Periodic boundary conditions are assumed, implying $r_{N+1} \equiv r_1$. The kinetic energy has not been included here. The adiabatic ground state potential of the two diabatic state Hamiltonian is [27]

$$V(r, R) = \frac{1}{2}(V_M(r) + V_M(R - r)) - \frac{1}{2}\sqrt{(V_M(r) - V_M(R - r))^2 + 4\Delta(R)^2} \tag{2}$$

with $\Delta(R) = \Delta_1 e^{-b(R-R_1)}$, $R_1 = 2r_0 + 1/a$ and $V_M(r)$ are Morse functions. The double-well potential experienced by a proton thus depends on the H-bond length R. Conversely, displacements of the proton from its equilibrium position exert a force on the oxygen atoms, providing a description of the coupling between proton positions and H-bond geometry characteristic for H-bonded ferroelectrics.

One Morse function is written as

$$V_M(r) = D\left(e^{-2a(r-r_0)} - 2e^{-a(r-r_0)}\right) \tag{3}$$

where r_0 is the equilibrium OH distance. The harmonic coupling for the oxygen-oxygen distance and between neighboring protons are written, respectively, as $V_{OO}(R) = \frac{1}{2}k_{OO}(R - R_{eq})^2$ and $V_{HH}(r_i, r_{i+1}) = \frac{1}{2}k_{HH}(r_i - r_{i+1})^2$. The parameters a, D, Δ_1, r_0, R_{eq}, k_{OO} and k_{HH} were fit to DFT calculations employing the vdW-DF2 functional, which was shown to describe squaric acid well in terms of H-bond geometry and proton transfer barrier [25]. The parameter b was set equal to the value of a, as suggested by McKenzie [27].

It should be noted that a similar model has been developed before to investigate the effect of quantum fluctuations on lattice constants at zero temperature [28], and McKenzie has investigated [29] isotope effects in the two diabatic state potential $V(r, R)$ used here.

2.2. DFT calculations

To fit the parameters, potential energy surfaces for proton transfer between neighboring molecules were calculated at various intermolecular O-O distances d_{OO} using DFT. Simply moving the O atoms on the neighboring squaric acid molecules closer to each other leads to unphysical structures. Instead, structures from the minimum energy path (MEP) were used since along this path d_{OO} varies from 2.590 Å (initial, geometry optimized structure) to 2.456 Å (transition state for proton transfer). The MEP was obtained from the nudged elastic band method [30]. For each (frozen) configuration of the C_4O_4 molecular frame along the MEP the potential energy surface for proton transfer was mapped out by transfering the proton along the O-O bond vector and performing DFT single-point calculations. The molecular frame distorts asymmetrically along the MEP when the proton is transfered from one oxygen to the next, but here symmetry is imposed by averaging the molecular structures on either side of the transition state (at fixed d_{OO}). Since the MEP involves a rotation of the C_4O_4 squaric unit, parametrizing the HBC model to the potential energy surface obtained in this manner will implicitly include this heavy-atom mode that couples to the proton transfer.

All DFT calculations were performed using the vdW-DF2 functional [26] in the plane-wave based VASP code [31, 32] with PAW potentials [33], an energy cutoff of 500 eV and a 4×1×4 k-point mesh. A simplified unit cell with only one $C_4O_4H_2$ molecule was used since interlayer interactions affect the proton-transfer barrier only slightly (less that 0.01 eV) [25].

2.3. Simulation methods

To investigate the ordering behavior of the HBC model it was implemented into the i-PI code [34] for MD and PIMD simulations. The PIMD simulations were run in the NVT ensemble using a white noise Langevin thermostat [35] in the normal mode representation and a local Langevin thermostat for the centroid coordinates, while a simple Langevin thermostat was used in MD. The required number of beads P was determined by increasing P until convergence was reached for quantum effects on geometrical distributions at a few temperatures. The number of beads to use at other temperatures was then determined by keeping the ratio β/P roughly constant, where $\beta = 1/k_BT$. For PIMD simulation of hydrogen the number of beads ranged from 240 at 50 K to 12 at 1000 K, and for deuterium P ranged from 84 at 100 K to 8 at 1000 K. A chain length of 300 H-bonds was used in all simulations.

3. Results

3.1. Parametrizing the HBC model

The right panel of Fig. 2(a) shows the symmetrized DFT-calculated double-well potentials for proton motion between the two oxygens in the H-bond, for different values of d_{OO}. The variable δ gives the position of the proton along the H-bond as $2\delta = d_{OO} - 2d_{OH}$ ($\delta = 0$ thus corresponds to the symmetric transition state). A single unit cell was used in the DFT calculations, with the potential energy surfaces thus corresponding to a collective transfer of all protons along a H-bonding chain. Figure 2(a), left panel, shows the best-fit model prediction obtained by

Table 1. Best-fit values of the parameters of the HBC model.

a (Å^{-1})	D (eV)	Δ_1 (eV)	r_0 (Å)	R_{eq} (Å)	k_{OO} (eV/Å^2)	k_{HH} (eV/Å^2)
2.7465	0.87558	0.18516	1.0613	2.650	8.00	3.40

a non-linear least-squares fit. Qualitatively the agreement with the DFT calculations is quite good, with the barrier height decreasing at smaller separations R.

Figure 2(b) shows DFT calculations and the model prediction for asynchronous proton transfer, *i.e.* where only one proton jumps in a 1x1x3 supercell. Introducing the harmonic coupling between protons in neighboring H-bonds, and thereby enforcing the ice rules, is seen to result in good agreement with DFT. The best-fit values of all parameters are given in Table 1. The parameters of the two-state diabatic potential determining the relevant energy scales for proton motion, D and Δ_1, are significantly smaller than the original parameters proposed by McKenzie [27], reflecting the small barrier for proton transfer in squaric acid as obtained by DFT/vdW-DF2.

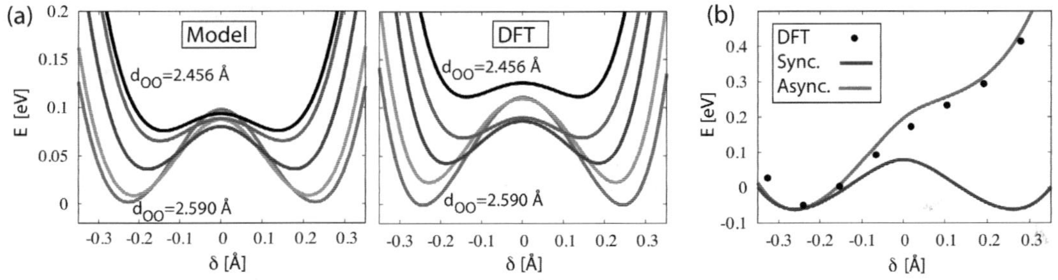

Figure 2. (a) Comparison of potential energy surfaces for synchronous jumps at different O-O distances d_{OO} with HBC model energy surfaces on the left and DFT energy surfaces on the right. (b) Potential energy surfaces for synchronous and asynchronous jumps for the HBC model compared to DFT energies for an asynchronous jump of one proton in a 1x1x3 supercell.

3.2. MD and PIMD simulations

In the following, results from an extensive series of simulations will be discussed. PIMD simulations have been performed using both hydrogen and deuterium masses (denoted by PIMD-H and PIMD-D) and MD simulations serve as classical reference points. Simulations are performed both using a fully flexible oxygen-oxygen distance d_{OO} and with a fixed value of d_{OO}, chosen here to be 2.6 Å.

3.2.1. Geometrical isotope effects The coupling between H motion and the H-bond length is built into the HBC model and one can thus use the PIMD simulations to investigate how the quantum delocalization of different H isotopes influences the H-bond geometry, *i.e.* the geometrical isotope effects. In our previous work [25] where squaric acid was simulated using ab initio PIMD we found these effects to be accurately reproduced with a lengthening of d_{OO}

by around 0.02 Å upon deuteration at room temperature. Figure 3 shows average d_{OO} for the flexible HBC simulations as function of temperature. A first key observation is the very large H/D isotope effect on d_{OO} which is largest at low temperatures and decreases at higher temperatures. A similarly large difference is seen between PIMD-D and classical MD (effectively infinite mass PIMD) simulations. These observations agree well with both experiments [36, 37] and our previous ab initio PIMD simulations [25], showing that the main physical mechanisms behind the geometrical isotope effect are captured by this simple model. Kinks can also be seen in the temperature dependence of d_{OO}, although the effect is hardly seen in PIMD-H simulations. As will be discussed below, these kinks indicate the presence of phase transitions related to the onset of proton disordering.

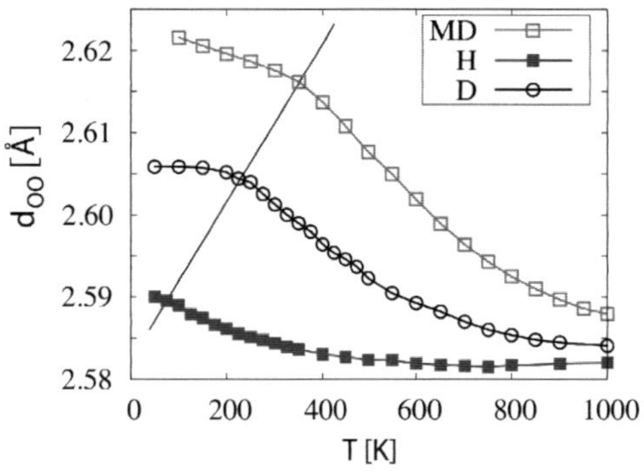

Figure 3. Average O-O distances d_{OO} as function of temperature where a linear correlation between T_C and d_{OO} is shown with a solid line.

3.2.2. Proton ordering in flexible-d_{OO} simulations I now turn to the analysis of proton ordering and begin by discussing the flexible-d_{OO} simulations. Figure 4 shows the average off-centering coordinate $\bar{\delta}$ and the value of the probability density $P(\delta)$ at the symmetric position $\delta = 0$, $P(\delta = 0)$, as function of temperature. $\bar{\delta}(T)$ values are extracted from the normalized probability densities $P(\delta)$ at each temperature. Similarly to the observation in Fig. 3, kinks are observed in the $\bar{\delta}(T)$ and $P(\delta = 0, T)$ curves. An inspection of the $P(\delta)$ probability densities confirms that these kinks are associated with the phase transition temperatures T_C above which collective jumps of H atoms become possible; $P(\delta)$ is unimodal below T_C and bimodal above T_C due to collective proton jumps. T_C is thus found at around 100 K, 250 K and 375 K for PIMD-H, PIMD-D and MD, respectively. The experimentally determined H/D isotope shift of T_C by 147 K is thus accurately reproduced, but the phase transitions occur at temperatures roughly 250 K below the experimental values, T_C(H)=373 K and T_C(D)=520 K. This is likely due to the proton transfer barrier, which may be underestimated in the DFT/vdW-DF2 calculations used to parametrize the HBC model. Indeed, the same downward shift of T_C was observed in our previous vdW-DF2 PIMD simulations [25] which explicitly accounted for all atomic degrees of freedom, where T_C(H) was located between 100 and 200 K and T_C(D) between 300 and 400 K.

Figure 4 demonstrates that neither the $\bar{\delta}$ coordinate nor the $P(\delta = 0)$ value alone determine the phase transition temperature, since their values differ significantly between PIMD-H, PIMD-D and MD at their respective transition temperatures of 100 K, 250 K and 375 K. On the other hand, nearly linear correlations are seen between T_C and both $\bar{\delta}$ and $P(\delta = 0)$.

A direct visualization of the ordering behavior is shown in Fig. 5 where the polarization of the H-bonded chains is plotted as function of simulation time. Each individual bond is represented on the vertical axis and pixels along horizontal lines are colored either gray or red according to the directionality of the bond, defined by which oxygen atom the H atom is closer to (for PIMD simulations the centroids of the ring-polymers are used). The top panels show the lowest

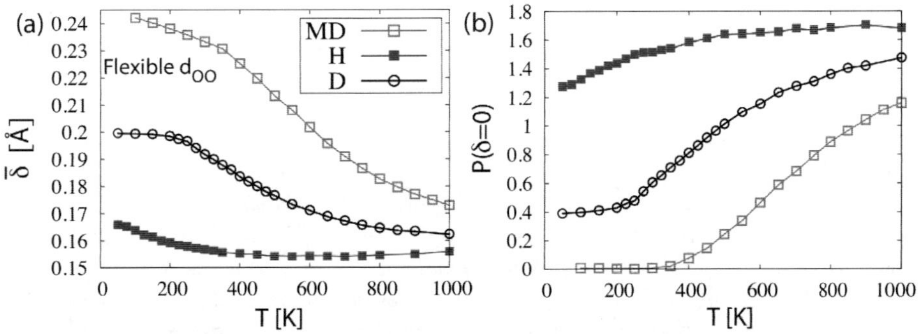

Figure 4. PIMD-H, PIMD-D and MD simulations with flexible-d_{OO}. (a) Mean value of the δ parameter as function of temperature. (b) Values of the $P(\delta)$ probability densities at δ=0 as function of temperature.

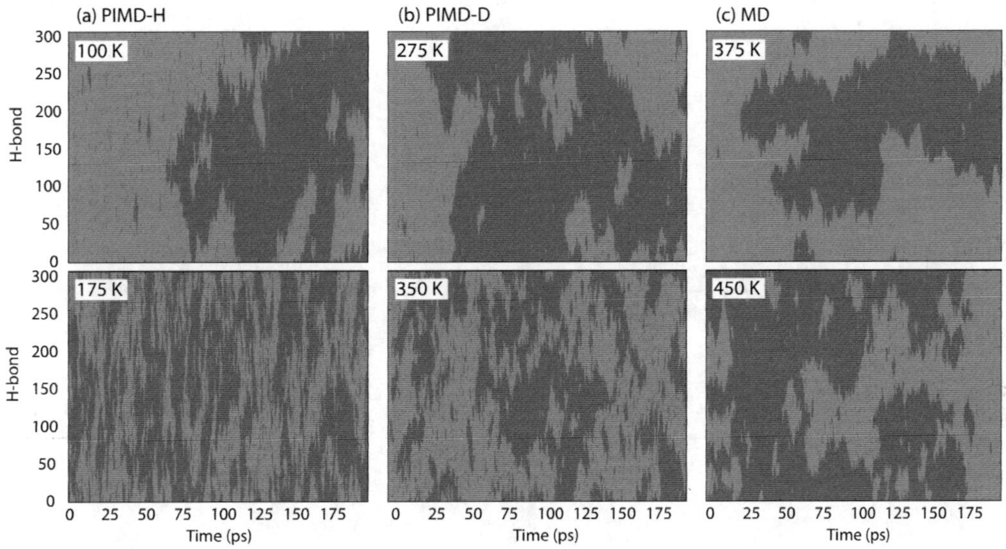

Figure 5. Maps of the polarization of the coupled H-bond chains as function of simulation time (horizontal axis) for each individual H-bond (vertical axis). Panel (a) shows PIMD-H simulations at 100 and 175 K, panel (b) shows PIMD-D at 275 and 350 K, and panel (c) shows MD simulations at 375 and 450 K.

temperatures, for PIMD-H, PIMD-D and MD simulations, below which ferroelectric ordering is observed and the lower panels show 75 K higher temperatures. Similar behavior is seen for the fixed-d_{OO} simulations. The large polarized regions in the upper panels of Fig. 5, *i.e.* just above the disordering phase transitions, reflect long-ranged spatial correlations along the H-bonding chains, in agreement with neutron scattering experiments on squaric acid [38] that indicated correlation lengths of at least 25 molecular units above T_C. At 75 K higher temperatures these correlations are of significantly shorter range. An isotope effect is clearly present in that the correlation length (size of polarized regions) is smaller for PIMD-H compared to PIMD-D at

$\sim T_C$+75 K, and smaller also for PIMD-D compared to classical MD. The sizes of correlated polarized regions and the dynamics and properties of domain walls will be investigated in future work.

3.2.3. Proton ordering in fixed-d_{OO} simulations Further insights into the nature of the phase transition can be obtained from the fixed-d_{OO} simulations, the results of which are shown in Fig. 6. As can be seen T_C shifts to higher temperatures for both MD and PIMD simulations, the kinks in $\bar{\delta}(T)$ and $P(\delta = 0, T)$ are now observed at around 200 K, 375 K and 425 K, for PIMD-H, PIMD-D and MD, respectively. A large isotope shift of T_C thus remains even though PIMD-H and PIMD-D simulations are performed at the same bond distance d_{OO}=2.6 Å. As discussed in the introduction, the exact role of quantum fluctuations in determining T_C has been debated since Blinc proposed [20] the proton-tunneling model for H-bonded ferroelectrics. One school of thought implicated the geometrical isotope effect as the major mechanism behind the T_C shift [37, 22]. However, the geometrical effect is inactive in the fixed-d_{OO} simulations and the observed shift of T_C must thus be directly related to the stronger quantum fluctuations of H compared to D atoms.

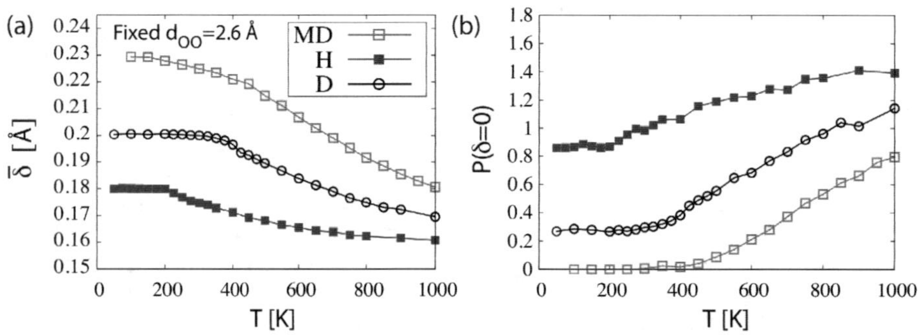

Figure 6. PIMD-H, PIMD-D and MD simulations with fixed-d_{OO}. (a) Mean value of the δ parameter as function of temperature. (b) Values of the $P(\delta)$ probability densities at δ=0 as function of temperature.

One difference between the fixed and flexible simulations concerns the extent of quantum delocalization of the two hydrogen isotopes. A standard measure of quantum delocalization is the radius of gyration, defined as $R_g^2 = \frac{1}{P}\sum_{i=1}^{P}(x_i - \bar{x})^2$ where P is the number of beads and $\bar{x}$ is the centroid coordinate of the ring-polymer. Figure 7 shows the average R_g as function of temperature for H and D atoms in PIMD where a comparison is made between the fixed and flexible-d_{OO} simulations. First, as expected, the H atoms are significantly more delocalized compared to D atoms with R_g(H) around 35% larger than R_g(D). On the other hand, while R_g in fixed and flexible simulations coincides at higher temperature, the increasing deviation of R_g between flexible and fixed-d_{OO} simulations at lower temperatures is interesting. An average flexible bond length shorter than a fixed bond length can be expected to correlate with larger quantum fluctuations since the H/D atom can more easily delocalize over the barrier, and this may partly explain why the flexible-d_{OO} PIMD-H simulations show larger R_g than the corresponding fixed-d_{OO} simulations. However, the increasing deviation in R_g at lower temperature contrasts with the decreasing difference between average d_{OO} in flexible PIMD-H and the frozen bond length d_{OO}=2.6 Å in the fixed-d_{OO} simulations (see Fig. 3). Furthermore, the average d_{OO} in the flexible PIMD-D simulations at low temperatures is larger than 2.6 Å,

Journal of Physics: Conference Series **571** (2014) 012012 doi:10.1088/1742-6596/571/1/012012

and yet R_g for flexible PIMD-D is larger than for the fixed PIMD-D at low temperatures. This behaviour is thus not simply a consequence of different average bond lengths. Instead, it can be understood as an effect where the O-O vibrational mode induces quantum delocalization of H/D atoms when the bond distance contracts, where quantum tunneling through a reduced energy barrier is the likely underlying mechanism. This effect has been called vibrationally enhanced tunneling [39].

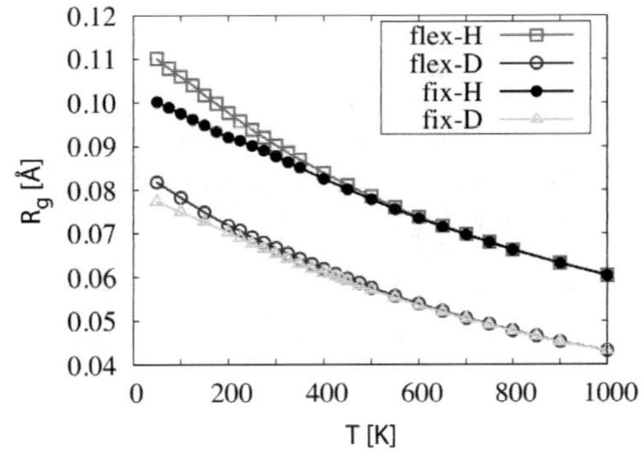

Figure 7. Radius of gyration R_g of H/D atoms as function of temperature for PIMD-H and PIMD-D in both flexible-d_{OO} and fixed-d_{OO} simulations.

4. Conclusions

In this work a simple potential model of a coupled H-bonded chain has been developed and parametrized to DFT calculations on squaric acid, a prototypical organic H-bonded ferroelectric. Through an extensive series of classical MD and quantum PIMD simulations I have illustrated that the model exhibits interesting properties that resemble the behavior of H-bonded ferroelectrics. The geometrical isotope effect at room temperature is reproduced in the simulations as an expansion by about 0.02 Å of H-bond distances upon deuteration. A proton-disordering ferroelectric-paraelectric phase transition occurs when collective jumps of protons across H-bonds become possible, and the transition temperature T_C is strongly dependent on the H atom mass; an H/D isotope shift of T_C of roughly 150 K is observed in good agreement with experiments on squaric acid. However, the transition temperatures for H and D are significantly lower than the corresponding experimental temperatures, which is likely due to an underestimated proton transfer energy barrier. Moreover, the simulated transition appears to be continuous while in experiments it is found to be of first order. This is likely because of the simplistic description in the model of the coupling to heavy atom degrees of freedom and the neglect of thermal expansion. Indeed, experiments indicate a mixed order-disorder and displacive character of the phase transition in squaric acid [40]. When fixing the O-O bond length d_{OO} in the model the phase transition temperature is shifted to roughly 100 K higher temperatures, even though the chosen value of d_{OO}=2.6 Å is smaller than the average d_{OO} at lower temperatures for PIMD-D simulations, and the H/D isotope shift of T_C remains. The latter observation shows that the geometrical isotope effect is not sufficient to explain the shift of T_C and, indeed, the quantum effects related to the different masses of H and D are crucial. The upwards shift of T_C after fixing d_{OO} can be understood from the enhanced quantum fluctuations of H and D atoms due to thermal vibrations of the H-bond distance, where shorter bond distances induce quantum delocalization and tunneling of the H/D atoms.

The initial simulation results reported here suggest that the H-bond chain model may prove useful for investigating in further detail the interplay between collective and quantum effects, the coupling between proton motion and lattice vibrational modes, and mechanisms behind disordering phase transitions in H-bonded ferroelectrics. Future work to improve the model will include refining the potential parameters to reproduce the correct T_C values, extending the model to include interchain correlations (both for parallel and perpendicular chains in the molecular planes and interplane interactions) and incorporating lattice thermal expansion. Interesting future applications will include investigating the properties of domain walls, determining critical exponents to compare with both experiments and Ising-type models, and to further elucidate quantum effects such as proton tunneling by computing momentum distributions and comparing to experimental deep inelastic neutron scattering measurements [41].

Acknowledgments

This research was supported by the Icelandic Research Fund through grant no. 120044042 and was conducted using the resources of High Performance Computing Center North (HPC2N) through the Swedish National Infrastructure for Computing (SNIC). The author would like to thank Simone Sturniolo and Ross McKenzie for useful discussions.

References

[1] Mathias G and Marx D 2007 *Proc. Natl. Acad. Sci. (USA)* **104** 6980–6985
[2] Brewer M L, Schmitt U W and Voth G A 2001 *Biophys. J.* **80** 1691–1702
[3] Kohanoff J J and Cachau R E 2004 *Mol. Phys.* **102** 1007–1014
[4] Dellago C, Naor M M and Hummer G 2003 *Phys. Rev. Lett.* **90** 105902
[5] Chen J, Li X Z, Zhang Q, Michaelides A and Wang E 2013 *Phys. Chem. Chem. Phys.* **15** 6344–6349
[6] Bove L E, Klotz S, Paciaroni A and Sacchetti F 2009 *Phys. Rev. Lett.* **103**(16) 165901
[7] Drechsel-Grau C and Marx D 2014 *Phys. Rev. Lett.* **112** 148302
[8] Horiuchi S, Kumai R and Tokura Y 2011 *Adv. Mat.* **23** 2098–2103
[9] Lee K, Kolb B, Thonhauser T, Vanderbilt D and Langreth D C 2012 *Phys. Rev. B* **86** 104102
[10] Valasek J 1921 *Phys. Rev.* **17** 475
[11] Blinc R 2002 *Ferroelectrics* **267** 3–22
[12] Schmidt V H 1987 *Ferroelectrics* **72** 157–173
[13] Koval S, Kohanoff J, Migoni R L and Tosatti E 2002 *Phys. Rev. Lett.* **89** 187602
[14] Horiuchi S and Tokura Y 2008 *Nature Mater.* **7** 357–366
[15] Horiuchi S, Tokunaga Y, Giovannetti G, Picozzi S, Itoh H, Shimano R, Kumai R and Tokura Y 2010 *Nature* **463** 789–792
[16] Mukhopadhyay S, Gutmann M J, Jura M, Jochym D B, Jimenez-Ruiz M, Sturniolo S, Refson K and Fernandez-Alonso F 2013 *Chem. Phys.* **427** 95–100
[17] Fernandez-Alonso F, Gutmann M J, Mukhopadhyay S, Jochym D B, Refson K, Jura M, Krzystyniak M, Jiménez-Ruiz M and Wagner A 2013 *J. Phys. Soc. Jap.* **82**
[18] Ehrhardt K D, Buchenau U, Samuelsen E J and Maier H D 1984 *Phys. Rev. B* **29** 996–1007
[19] Palomar J and Dalal N S 2002 *Ferroelectrics* **272** 173–179
[20] Blinc R 1960 *J. Phys. Chem. Solids* **13** 204–211
[21] Li X Z, Walker B and Michaelides A 2011 *Proc. Natl. Acad. Sci. (USA)* **108** 6369
[22] McMahon M I, Nelmes R J, Kuhst W F, Dorwarth R, Piltz R O and Tun Z 1990 *Nature* **348** 317–319
[23] Reiter G F, Mayers J and Platzman P 2002 *Phys. Rev. Lett.* **89** 135505
[24] Koval S, Kohanoff J, Lasave J, Colizzi G and Migoni R L 2005 *Phys. Rev. B* **71** 184102
[25] Wikfeldt K T and Michaelides A 2014 *J. Chem. Phys.* **140** 041103
[26] Lee K, Murray É D, Kong L, Lundqvist B I and Langreth D C 2010 *Phys. Rev. B* **82** 081101
[27] McKenzie R H 2012 *Chem. Phys. Lett.* **535** 196–200
[28] Scivetti I, Gidopoulos N and Kohanoff J 2008 *Phys. Rev. B* **78** 224108
[29] McKenzie R H, Bekker C, Athokpam B and Ramesh S G 2014 *J. Chem. Phys* **140** 174508
[30] Henkelman G, Uberuaga B P and Jónsson H 2000 *J. Chem. Phys.* **113** 9901
[31] Kresse G and Furthmüller J 1996 *J. Comp. Mater. Sci.* **6** 15–50
[32] Kresse G and Furthmüller J 1996 *Phys. Rev. B* **54** 11169–11186
[33] Kresse G and Joubert D 1999 *Phys. Rev. B* **59** 1758
[34] Ceriotti M, More J and Manolopoulos D E 2013 *Comp. Phys. Comm.* **185** 1019
[35] Ceriotti M, Parrinello M, Markland T E and Manolopoulos D E 2010 *J. Chem. Phys.* **133** 124104
[36] Semmingsen D, Hollander F J and Koetzle T F 1977 *J. Chem. Phys.* **66** 4405
[37] McMahon M I, Nelmes R J, Kuhs W F and Semmingsen D 1991 *Z. f. Kristallogr.* **195** 231–239
[38] Samuelsen E J and Semmingsen D 1977 *J. Phys. Chem. Solids* **38** 1275–1283

[39] Bruno W J and Bialek W 1992 *Biophys. J.* **63** 689–699
[40] Dalal N, Klymachyov A and Bussmann-Holder A 1998 *Phys. Rev. Lett.* **81** 5924
[41] Andreani C, Colognesi D, Mayers J, Reiter G F and Senesi R 2005 *Adv. Phys.* **54** 377–469

Journal of Physics: Conference Series **571** (2014) 012013 doi:10.1088/1742-6596/571/1/012013

Discussion: Theoretical Horizons and Calculation

M. Ceriotti,[1], C. Drechsel-Grau,[2] F. Fernandez-Alonso,[3,4] N. Greaves,[5,6] M. Krzystyniak,[3,7] D. Major,[8] A. Parmentier, [10] C. Salzmann,[9] R. Senesi,[10,11] S. Sturniolo,[3] L. Walewski,[2] T. Wikfeltd.[12]

[1] Laboratory of Computational Science and Modelling, NQ Institute of Materials, École Polytechnique Fédérale de Lausanne, CH-1015 Lausanne, Switzerland
[2] Lehrstuhl für Theoretische Chemie, Ruhr-Universität Bochum, 44780 Bochum, Germany
[3] ISIS Facility, Rutherford Appleton Laboratory, Chilton, Didcot, Oxfordshire OX11 0QX, United Kingdom
[4] Department of Physics and Astronomy, University College London, Gower Street, London WC1E 6BT, United Kingdom
[5] Centre for Advanced Functional Materials and Devices, Institute of Mathematics and Physics, Aberystwyth University, Aberystwyth SY23 3BZ, United Kingdom
[6] Department of Materials Science and Metallurgy, University of Cambridge, Cambridge BB2 3QZ, United Kingdom
[7] School of Science and Technology, Nottingham Trent University, Clifton Campus, Nottingham NG11 8NS, United Kingdom
[8] Chemistry Department, Bar-Ilan University, Raman-Gan 5290002, Israel
[9] Department of Chemistry, University College London, 20 Gordon Street, London WC1H 0AJ, United Kingdom
[10] Università degli Studi di Roma Tor Vegata, Dipartimento di Fisica and Centro NAST, via della Ricerca Scientifica 1, Rome 00133, Italy
[11] CNR-IPCF Sezione di Messina, Messina, Italy
[12] Science Institute, University of Iceland, Iceland

1. Inexpensive Evaluation of the Quantum Kinetic Energy and Particle Momentum Distribution by First-Principles Simulations (Michele Ceriotti, EPFL, Switzerland)

Dr Walewski opened the discussion of the lecture by Prof Ceriotti: While calculating the anisotropic kinetic energy tensor by running the sliding window average, one looses the rigorous canonical ensemble average property. Would it be possible to perform the average in a microscopic internal coordinate system relative to the molecular frame?
Prof Ceriotti replied: Yes, it can be done, either by defining a physically-motivated reference frame or by aligning molecules to a fixed reference and rotating the kinetic energy tensor estimator accordingly before averaging. The problem with these approaches is that they require one to have well-defined and relatively rigid molecular entities. For instance, this would not work in the presence of excess protons. As shown in Romanelli, G., Ceriotti, M., Manolopoulos, D. E., Pantalei, C., Senesi, R., and Andreani, C. (2013). 'Direct Measurement of Competing Quantum Effects on the Kinetic Energy of Heavy Water upon Melting' *J. Phys. Chem. Lett.* **4**(19), 3251-3256., there is only a 1% difference between the result obtained with a moving average and that obtained by aligning the molecules to a fixed reference.

Journal of Physics: Conference Series **571** (2014) 012013 doi:10.1088/1742-6596/571/1/012013

Prof Greaves communicated: Is it possible for you to simulate the amorphization of hexagonal ice to high-density water?
Prof Ceriotti communicated in response: We have considered this, however it is always hard to build reliable computational models of meta-stable phases, requiring a complex preparation protocol that cannot be precisely mimicked in simulations. In particular, given the time scales involved, one would have to generate the model using an empirical water model and then run an ab initio simulation to extract PMD data for that configuration. It is tricky, but it would certainly be an interesting problem to study theoretically.

Dr Krzystyniak queried: There's a relation between the coefficients of Gram-Charlier expansion and multivariate $n(p)$. Why choosing the multivariate Gaussian model?
Prof Ceriotti replied: It is easy to approximate $n(p)$ as a multi-variate Gaussian distribution because one can compute the second moments of $n(p)$ from a conventional path integral simulation. To obtain the actual PMD one needs to perform an open-path simulation, which is much more demanding.

2. Collective Proton Tunneling in Ordinary Ice (Christoph Drechsel-Grau, Bochum, Germany)

Dr Salzmann opened the discussion of the lecture by Dr Drechsel-Grau: Statistically speaking, the chance of defect migration along a six-membered ring is very small. From your results, would you argue that this is still preferred over migrations over larger rings or even chains?
Dr Drechsel-Grau replied: We used constraints that prevent protons from exiting the six-membered ring.

Dr Sturniolo communicated: Is there any estimate about the relationship between single-proton and collective-proton tunneling probabilities?
Dr Drechsel-Grau communicated in response: No, since single jumps do not occur in a stable system.

3. Simulation Studies of Hydrogen Bonded Ferroelectrics (Thor Wikfeldt, Reykjavik/Stockholm)

Dr Ceriotti opened the discussion of the lecture by Dr Wikfeldt: The gyration radius of a ring polymer is not a physical observable. How could we describe delocalization in a more physically motivated way? For instance, in Parrinello, M., and Rahman, A., 'Study of an F center in molten KCl' *J. Chem. Phys.*, **80**(2), 860-867 (1984) they define an estimator for the spread of the density associated with a single particle.
Dr Wikfeldt replied: It is true that an increase in a particle's radius of gyration in PIMD simulations is not direct proof that quantum tunnelling is at play. However, a strong delocalization of the ring-polymer along a reaction coordinate, where the spread resembles an instaton path, is a strong indicator of tunnelling. This is what we observe in our work on squaric acid (*J. Chem. Phys.*, **140**, 041103 (2014)).

Journal of Physics: Conference Series **571** (2014) 012013 doi:10.1088/1742-6596/571/1/012013

4. Interaction Induced Localization of Protons: Temperature and Superfluid Helium Solvation Effects (Lukasz Walewski, Bochum, Germany)

Prof Fernandez-Alonso opened the discussion of the lecture by Dr Walewski: What is the correlation between localization index in the gas phase and intermolecular He-H, He-O, He-Cl interactions?
Dr Walewski replied: This still needs to be investigated.

Prof Fernandez-Alonso communicated: Can H_2 be treated by your computational method?
Dr Walewski communicated in response: Yes, for some specific systems .

Dr Senesi commented: Is there any estimate of the excess kinetic energy due to the interaction induced by the localization of protons?
Dr Walewski replied: This will be calculated in the near future.

5. Nuclear Quantum Effects in Enzyme Catalysis from Momentum Distributions (Dan Major, BIU, Israel)

Prof Ceriotti opened the discussion of the lecture by Prof Major: Was any bias added on ring polymer centroids or the individual beads?
Prof Major replied: There was no added bias.

Journal of Physics: Conference Series **571** (2014) 012014 doi:10.1088/1742-6596/571/1/012014

Over the Horizon: Future Roles of Electron Volt Neutron Spectroscopy

A.G. Seel[1,2], M. Krzystyniak[1,3], R. Senesi[4,5], and F. Fernandez-Alonso[1,6]

[1] ISIS Facility, Rutherford Appleton Laboratory, Chilton, Didcot, Oxfordshire OX11 0QX, United Kingdom

[2] Department of Chemistry, University of Oxford, South Park Road, Oxford OX1 3QR, United Kingdom

[3] School of Science and Technology, Nottingham Trent University, Clifton Campus, Nottingham NG11 8NS, United Kingdom

[4] Università degli Studi di Roma Tor Vergata, Dipartimento di Fisica and Centro NAST, via della Ricerca Scientifica 1, Rome 00133, Italy

[5] CNR-IPCF Sezione di Messina, Messina, Italy

[6] Department of Physics and Astronomy, University College London, Gower Street, London WC1E 6BT, United Kingdom

The final session of the *VI Workshop on Electron Volt Spectroscopy* was devoted to an open discussion of the future roles of the technique, taking as starting point the results and perspectives presented and discussed over this two-day event. In this context, it is recognised that VESUVIO at ISIS has already been successful at providing unique insights into Nuclear Quantum Dynamics (NQD) of light nuclides in condensed matter, so far with an emphasis on model systems that provide precise and much-needed benchmarks for state-of-the-art theoretical predictions. More recent infrastructure and scientific developments have also shown the yet-to-be-tapped potential of MAss-selective Neutron SpEctroscopy (MANSE) to study functional materials, a convenient gateway to expand the core capabilities of the Deep Inelastic Neutron Scattering (DINS) technique beyond its current remit.

To conclude these proceedings, we provide a summary of ideas and perspectives brought up during this final discussion session, emphasizing the importance of joint experimental and theoretical efforts to understand and describe NQD in condensed matter, growing synergies across a number of scientific disciplines, and ongoing and further developments in novel instrumentation across the globe.

The Evolving State of DINS

What became immediately clear from this meeting is that DINS has been in a state of evolution from its very inception, representing a technique that is unique to spallation neutron sources that produce a suitable flux of epithermal neutrons. The VESUVIO spectrometer has been in operation at ISIS in one form or another since 1989 as its initial incarnation, EVS [1]. The user programme on EVS soon produced experiments examining NDQ of protons in condensed matter, culminating in the first single-crystal measurements of momentum distributions in potassium di-hydrogen phosphate [2]. This represented a stark departure from previous applications of DINS, which centred around helium and the associated macroscopic phenomena of superfluidity and

Journal of Physics: Conference Series **571** (2014) 012014 doi:10.1088/1742-6596/571/1/012014

Bose-Einstein condensation [3]. DINS was then at the level whereby, in favourable circumstances regarding sample availability, the effective single-particle potential energy surface for protons in condensed matter could be accessed experimentally.

Continued efforts were made to upgrade instrumentation, with the VESUVIO (2004) and eVERDI (2008) projects being the major milestones towards the instrument as it stands today. The current user programme on VESUVIO has witnessed another evolution in recent years, towards the investigation of functional materials, and those of increasingly complex nature. Our discussion session allowed us to clarify where we are in terms of research themes, the manner in which data is both collected and analysed, and the increasingly important role of computation and materials modelling in the future of DINS.

NQD and MANSE: Fundamental Systems and Beyond

Nuclear quantum effects underpin much of the current research in the field of DINS. In addition to systems of fundamental interest, such as water (as represented in this meeting by the work of Andreani, Reiter, Romanelli, Drechsel-Grau, and Ceriotti), there is increasing interest in functional materials that demonstrate a distinctly non-harmonic proton environment (Wikfeldt and Krzystyniak). DINS thus has a place within the fields of applied materials research, such as proton conductors, organic piezo- and ferroelectrics and ionic liquids. It is particularly worthy to note that, through the VESUVIO user programme, interest has been shown recently for all of these systems.

This meeting has also brought to the fore the more recent emergence of MANSE. In the preceding sessions on this topic, we were presented with light nuclides that are of interest in both fundamental and applied science (such as Li, C, N, O) and accessible for currently operational DINS spectrometers such as VESUVIO. Furthermore, Krzystyniak presented additional possibilities to extract mean kinetic energies for heavier nuclei via the use of resonant Dopplerimetry.

Both NQD and MANSE are areas in continual development, both in terms of experimental and computational techniques, and the interplay between them. As such, the results of our discussion session are presented below so as to highlight the next steps in examining NQD in condensed matter, as driven by an increasing need to routinely examine complex materials in which they are exhibited, and the complementary MANSE technique.

[1] **Emerging experimental requirements.** The first question we must ask is just how sensitive is DINS to the physico-chemical environment of the proton, or any given mass, and to intrinsic nuclear quantum effects. To first approximation, by how much should we expect the mass-specific Compton width, σ_M, to vary in a given system? The long measurement time on VESUVIO has limited the number of truly parametric studies that have been conducted to date, and this limitation was immediately highlighted in this discussion session. An expedient 'next step' for this field would be to increase the number of studies conducted over varying temperatures, pressures, or chemical trends. Such parametric studies would not only be informative in terms of experimental sensitivity, giving a bounded region for the environmental sensitivity for the proton and heavier masses, but also act as a benchmark for computational studies that are capable of simulating experimental conditions.

Further to the above, discussion centred on modern measurements of systems such as $H_2/D_2/HD$ where NQD phenomena such as particle exchange can be effectively 'switched on/off.' A systematic overview of these systems would enable quantification of our current experimental sensitivity to specific NQD effects. In this context, important too is the selection of samples for investigation in terms of availability, stability, and form. Access to single-crystal samples, large enough for DINS measurements, is the only route towards direct measurement of truly multidimensional (in a non-spherically-averaged sense) proton Compton profiles, kinetic energies, and eventually potential energy surfaces.

Journal of Physics: Conference Series **571** (2014) 012014 doi:10.1088/1742-6596/571/1/012014

[2] **Emerging conceptual, theoretical and computational needs.** The most significant factor in the modelling and interpretation of NQD has been the move beyond the harmonic approximation and periodic media. This lauded approximation still represents the cheapest way to calculate zero-point energies, and is still the workhorse for the interpretation of neutron spectroscopy probing atomic vibrations at lower energy transfers. The aim of computation as applied to spectroscopic interpretation or modelling is to establish well-defined protocols to explore NQD at various levels of controlled and hierarchical approximation. This task can be achieved via recourse to classical molecular dynamics (exhibiting no NQD), harmonic calculations (as a starting point to include or identify dominant quantum effects such as zero-point-energy motions), quasiharmonic and anharmonic corrections including increasinly popular self-consistent approaches, and through to fully quantum-mechanical methodologies which necessarily need to include particle delocalisation, interference, and exchange in condensed-matter media. In this manner, these computational tools would be available for any given system measured by DINS, from the more classical end and the overlap with MANSE, through to the fully quantum nature of proton dynamics in a plethora of situations of both fundamental and practical interest.
Experimental design could be greatly aided through a combination of (the more widespread) phonon calculations (primarily using first-principles electronic structure codes, generally DFT), as well as PIMD methods (as highlighted in this meeting by the i-Py programme of Ceriotti *et al* [4]). The more widespread use of these tools would allow for the identification of systems possibly exhibiting pronounced NQD, or wherein nuclear dynamics show a strong variance with temperature, as well as for the subsequent interpretation of experimental data. At this point, it is important to address a recurring theme arising from this discussion session, namely, the link between DINS experiments and PIMD calculations. Of particular importance is the study of protons and the manner in which the momentum distribution $n(p)$ is interpreted. On the one hand, the experimental DINS community have traditionally used a Gram-Charlier expansion to fit measured neutron Compton profiles and yield the isotropic $n(p)$. On the other hand, a multivariate Gaussian model is more conducive to physical interpretation and comparison with PIMD. It is therefore important to reach a future synergy between computation and experimental interpretation for the investigation of NQD, particularly in isotropic systems. This must also include deviations from the 'ideal response' under the celebrated impulse approximation, also known as final-state effects.

[3] **Systems and frontiers for DINS.** Our discussion on NDQ and MANSE ended with the suggestion of systems that could (and should) be highlighted for further experimental and theoretical scrutiny. Included too are their associated challenges. Chief amongst these were complex hydrogen-bonding systems, both natural and man-made, for the study of NQD via DINS. An example field mentioned was mineralogy, where the proton response in a series of chemically related, hydrogen-bonded frameworks could be determined. This would enable a direct link between the observables of DINS (such as σ_M) and chemical binding parameters such as bond strength, acidity/basicity, or toughness. This area is also of interest for the future development of MANSE, where it was recognised that this 'parametric link' is a crucial step to open up the field of DINS and NQD to the wider chemistry and materials communities.
Selective deuteration was also commented upon as crucial for future investigations of proton-containing systems. This is particularly true for increasinly complex organic materials that comprise distinctly different proton environments, with discernibly different NQD at each site (accessible through computation). The ability of DINS to isolate the proton from its heavier isotopic analogue the deuteron (and to measure both simultaneously) renders it a

unique tool for NQD studies in organic materials. This mass-specific isolation of the proton is also not affected by temperature in the same way as the Debye-Waller factor in diffraction or inelastic neutron scattering. These factors are particularly important were the field of DINS to evolve in the direction of biological processes, whereby NQD plays a central role in catalytic processes in biology (as discussed in the presentation of Major), provided that selective deuteration is available to isolate relevant protons.
Other areas that were highlighted include interfacial phenomena, reduced-dimensionality materials (exemplified by graphitic and Li-conducting materials) and industrially important observables such as H/D/T levels in materials for fusion technology (in the included contribution from Dawidowski *et al*). For reduced dimensionality samples, whilst single-crystals may not be available, we can often find partially orientated (or stratigraphically disordered) samples that can be aligned relative to the incident neutron beam. Two-dimensional momentum distributions can thus be accessed experimentally. For protons at interfaces and defects, it was pointed out that the main limitation for these systems lies in the potentially low concentration of protons at these sites, and that the absolute sensitivity for current instrumentation is yet to be ascertained in depth. Nevertheless, a lower bound for proton concentration has already been estimated experimentally, as discussed below.

Instrumentation and Data Analysis Protocols

Whereas our discussion was intended to highlight areas of research or specific systems that we could approach with DINS, there occurred throughout (from both experimentalists and theorists) comment and clarification for the future developments in instrumentation and data analysis. As such, we have collected here the most significant ways in which both instrumentation and data analysis could be developed in line with the user community and science cases highlighted above.

[1] **Useful flux and epithermal neutrons.** The primary limitation of current instrumentation, particularly in inverted-geometry as in the case of VESUVIO, is flux (commented upon by Seel and Senesi). This is both a problem of available and detected flux, the former being how many epithermal neutrons irradiate the sample, the latter reflecting how many are counted. Any increase in available flux is reliant upon the target station to which the spectrometer is attached (in the case of VESUVIO this would be ISIS Target Station I), and so is in a sense instrument independent. Nevertheless, an increase of a factor of 5 in incident flux would revolutionise the manner in which DINS measurements are conducted, particularly in relation to parametric studies as highlighted above. In terms of detected flux, there are a number of ways the effective count-rate for VESUVIO in particular can be increased, not least of which is an increase in the number of detectors, particularly in backscattering. Included in discussion towards this goal was a decrease in detector size or synchronous measurement with several analyser foils. Further to this was a discussion on the manner in which we analyse data, be it on a detector-by-detector basis, groupings of detectors to effectively reduce signal-to-noise and increase count-rate (Seel), or using cumulative strategies aimed at a determination of specific experimental observables such as mean kinetic energies within some level of confidence (Krzystyniak).

[2] **Improving instrument resolution.** Resolution is an important factor when studying NDQ in detail or in terms of mass-isolation in MANSE. As commented on by Seel and Senesi, this can be done in a number of ways through the alteration and redesign of detector-bank geometry. The angular component to proton resolution on VESUVIO could be vastly improved by redesigning the forward-scattering detector banks, with an energy resolution dependent upon analyser foils and foil cycling (Gorini). The manner in which the isolation and separation of light-mass components could be improved was commented upon (see contributions to these proceedings by Krzystyniak and Seel), always dependent upon an increase in incident neutron flux. The resolution of the instrument

Journal of Physics: Conference Series **571** (2014) 012014 doi:10.1088/1742-6596/571/1/012014

for resonant Dopplerimetry (applicable to heavy masses) was also discussed, in terms of possible extensions to existing detector technologies for finer resolution and improved count rates.

[3] **Establishing element-specific parameters.** As discussion on the typical duration of experiments progressed, particularly in light of MANSE, it became apparent that the absolute sensitivity of current instrumentation is not well defined. For a given sample (including proton-containing systems), a guide to sample size and measurement duration regarding spectral statistics should be benchmarked. For the proton (and carbon), an experiment has recently been conducted on various thicknesses of polyethene, but for systems containing any mass accessible in the DINS regime, particularly lithium, a lower-bound to mass-specific concentration needs to be determined. For complex systems containing many masses (as presented by Greaves and Chass) it would be of great help for users to have access to a spectral simulator, whereby a sample stoichiometry is defined and an idealised MANSE spectra in time of flight simulated. This has recently been implemented within the MANTID framework, but does not as yet take into account instrumental noise levels.

[4] **Sample environment and ancillary measurements.** For DINS measurements beyond the lightest masses, contributions to spectra from sample environment often complicates analysis. This is particularly true due to long measurement times rendering equal time on background measurement unfeasible in many cases. To this end, discussion on manners to improve upon existing sample environment turned to the use of aluminium as a sample container and instrument material. By verity of its mass (27 amu), aluminium signals always overlap with any mass greater than lithium. As pointed out by Seel, a move towards thinner sample containers of a heavy metal, such as tin, would render spectral isolation of sample environment far simpler. A prototype CCR that includes tin rather than aluminium windows is currently being produced in order to further enlarge upon this concept
Often overlooked is the existing ability of DINS spectrometers to act as diffractometers, particularly those operating in inverted geometry as with VESUVIO, where resolution in the thermal region in backscattering is particularly good. This ability to simultaneously measure diffraction and DINS has already been utilised in the study of layered materials [5]. Furthermore, by including analyser foils in the incident flight path, it was pointed out during this discussion session that the capability of the instrument to measure vibrational data (albeit with relatively poor resolution) could be enhanced. Beyond DINS spectrometers, this could be tested on instruments with similar incident energy profiles, such as SXD at ISIS [6]. As pointed out by Dawidowski in these proceedings, these also exists great potential for the measurement of total neutron scattering cross sections via the operation of instruments such as VESUVIO in transmission mode. DINS spectrometers operate in an energy region that is poorly quantified in comparison to those both higher and lower in incident neutron energy, and should be of interest in the field of nuclear engineering. This capability would act as a natural partner to the resonance analysis discussion of Krzystyniak.

[5] **Data analysis and mining.** The manner in which users view, access and can analyse DINS data from VESUVIO is currently being standardised under the umbrella of the MANTID framework, as illustrated within these proceedings by Jackson *et al.* Existing capabilities including the handling of raw data, data reduction and correction (for both γ-background and multiple scattering effects) and data fitting will soon be routinely available. This was met with a favourable response from experimentalists within this discussion session, with a further call to enhance existing capabilities to include more robust statistical methods, such as Bayesian inference or principal-component analysis. The sparse nature of DINS spectra was highlighted as the driving force for these initiatives, and these statistical techniques are currently being examined for implementation within MANTID. Of equal interest from the

computational members of this session was the possibility of moving beyond an experimental probe of atomic momentum distributions and kinetic energies, into a standardised mean-force determination and future emphasis on the exploitation of final-state effects (those 'non-ideal' parameters mentioned above) including their overlap with theory.

It was clear from these discussions that the manner in which we treat and analyse DINS data has evolved over the years (see contributions from Romanelli, Senesi and Reiter) and that there exists much measured data that could be reanalysed in light of this and the capabilities of modern computational methods. Such 'data mining' has the possibility to quickly extend the interplay between experiment and computation without reliance on future measurements alone. It would also be possible to produce a current database of measured and computed kinetic energy values available to the community which would underpin the position of DINS within a wider scientific context. Furthermore, as more instruments capable of DINS measurements become available there will be an increased need to standardise the manner in which data is presented. This was mentioned in light of future MANSE developments, where data is often presented in time-of-flight and is thus instrument dependent. Were this scale to be transformed to a standardised 'shift,' future data representation could be rendered instrument-independent in the same manner as techniques such as NMR (chemical shift) or diffraction (d-spacing).

Quo Vadis DINS

It became clear from this intense and productive two-and-a-half-hour discussion session, and when compiling these proceedings, that there already exists a strong and growing user community for DINS, both in terms of experimentalists and computational researchers. Representatives from the fields of physics, chemistry, applied materials and biological research each had some input in the current and future role of DINS as a spectroscopic technique utilising electron-volt neutrons. In particular, we must remind ourselves what renders DINS unique, special and exciting; something that we believe can be divided into two complementary manners of viewing DINS measurements. On the one hand we have a 'quantum diffractometer' - a technique that can access nuclear probability distributions through detailed analysis of DINS line shapes. On the other hand we have an 'atomic quantum thermometer' - a mass-specific technique able to determine atomic kinetic energies as a direct measure of chemical environment. Admittedly, these are somewhat simplified views of the power of DINS, the full potential of which is yet to be reached. The *VI Workshop on Electron Volt Spectroscopy* has shown that this field is continuing to develop along new pathways, and in a likewise fashion we hope to see the potential of DINS grow as we progress towards the next meeting in this series.

Acknowledgments

This work was supported within the CNR-STFC Agreement No. 06/20018 concerning collaboration in scientific research at the ISIS Pulsed Neutron & Muon Source. A.G.S., M.K. and F.F.A. gratefully acknowledge the UK Science and Technology Facilities Council for financial support.

References

[1] Mayers J and Reiter G 2012 *Meas. Sci. Technol.* **23** 045902
[2] Reiter GF, Mayers J and Platzman P 2002 *Phys. Rev. Lett.* **89** 135505
[3] Andreani C, Colognesi D, Mayers J, Reiter GF and Senesi R, 2005 *Adv. Physics* **54** 377-469
[4] Ceriotti M, More J and Manolopolous DE 2014 *Comput. Phys. Commun.* **185** 1019-1026
[5] Krzystyniak M, Adams MA, Lovell A, Skipper NT, Bennington SM, Mayers M and Fernandez-Alonso F 2011 *Faraday Discuss.* **151** 171-197.
[6] Keen DA, Gutmann MJ and Wilson CC 2006 *J. Appl. Cryst.* **39** 714-722

AUTHOR INDEX

Institute of Physics
Dirac House, Temple Back
Bristol BS1 6BE UK

CURRAN ASSOCIATES INC
proceedings
.com